I0470165

U.S. Fire Administration

Traffic Incident Management Systems

FA-330/March 2012

FEMA

U.S. Fire Administration

Mission Statement

We provide National leadership to foster a solid foundation for our fire and emergency services stakeholders in prevention, preparedness, and response.

Table of Contents

Preface

The U.S. Fire Administration (USFA) would like to acknowledge the U.S. Department of Transportation (DOT) Federal Highway Administration (FHWA) for its support of this project. Several members of the FHWA staff also served as reviewers of this report, including Emergency Transportation Operations Team Leader Kimberly C. Vasconez and Tim Lane.

This report was developed through a cooperative agreement between the USFA and the International Fire Service Training Association (IFSTA) at Oklahoma State University (OSU). IFSTA and its partner OSU Fire Protection Publications has been a major publisher of fire service training materials since 1934. Through its association with the OSU College of Engineering, Architecture, and Technology it also conducts a variety of funded, technical research on fire service, fire prevention, and life safety issues.

The extensive information provided within this report would not have been possible without the dedication and efforts of the following people assigned to this project:

- Project Administrator—Nancy Trench, Assistant Director for Research, OSU Fire Protection Publications;

- Principle Investigator/Editor—Michael A. Wieder, Executive Director, IFSTA; and

- Document Development—Ben Brock, Senior Graphic Designer, OSU Fire Protection Publications.

The USFA would also like to acknowledge the efforts of the National Fire Service Incident Management Consortium in developing the excellent procedures for applying the Incident Command System (ICS) to highway incidents that are outlined in this document. This information was excerpted from the Consortium's "IMS Model Procedures Guide for Highway Incidents" that was developed with funding from the DOT. We are grateful for the use of that information in this report.

Chapter 1 Introduction

In 2003, the U.S. Fire Administration (USFA) announced a goal to reduce firefighter fatalities by 25 percent within 5 years and 50 percent within 10 years. It also committed to doing research that would support that goal. The consistently high annual percentage of fatalities related to fire department response and roadway scene operations prompted the USFA to look at several aspects related to these collisions in an effort to improve responder safety.

Firefighters who are killed in privately owned vehicles (POVs) during the course of their duties account for the largest percentage of vehicle-related deaths. These are typically volunteer firefighters who are responding to or returning from emergency calls. However, career firefighters are also occasionally killed in POVs while performing their duties.

One of the USFA's initial forays into the responder roadway safety issue was through its cooperative work with the Cumberland Valley Volunteer Firemen's Association (CVVFA) and its Emergency Responder Safety Institute (ERSI) and ResponderSafety.com website. The CVVFA is an association of individual and organizational fire service members from the mid-Atlantic region of the United States. It is very active in a variety of fire service issues. It was one of the leaders in identifying the need for improved methods to protect emergency responders who are operating at roadway incidents. It has participated in numerous interagency and multidisciplinary projects related to this issue. It has also developed some of its own training packets, such as the "Slow Down and Move Over" public service announcement (PSA) to spread the message about the dangers of working on the roadway. For more information on the CVVFA and its roadway safety programs go to: www.cvvfa.org

Figure 1.1. Many firefighters are injured or killed as a result of apparatus collisions. *Courtesy of Ron Jeffers, Union City, NJ.*

Fire department tankers (tenders) account for the most firefighter response-related fatalities in fire apparatus. More firefighters are killed in tankers than in pumpers and ladder apparatus combined. In response to the alarming numbers of fatalities occurring in tankers, the USFA published "Safe Operations of Fire Tankers" (FA-248) in 2003.

In partnership with the U.S. Department of Transportation (DOT)/National Highway Transportation Safety Administration (NHTSA) and the DOT/Intelligent Transportation Systems (ITS) Joint Program Office (JPO), USFA initiated the Emergency Vehicle Safety Initiative (EVSI) in 2002. The initiative:

- identified the major issues related to firefighter fatalities that occur while responding to or returning from alarms and while operating on roadway emergency scenes (Figure 1.1); and

- developed and obtained consensus among major national-level fire and emergency service trade associations on draft "best practices" guidelines, mitigation techniques, and technologies to reduce firefighter response and roadway scene fatalities.

Emergency Vehicle
Safety Initiative

FA-272 / August 2004

FEMA

Figure 1.2. The Emergency Vehicle Safety Initiative was released by the USFA in 2004.

The USFA published the results in "Emergency Vehicle Safety Initiative" (FA-272) in August 2004 (Figure 1.2). The report identified several recommendations to improve safety related to response and highway operations.

As a followup to the "Emergency Vehicle Safety Initiative," USFA initiated partnerships with the International Association of Fire Chiefs (IAFC), the International Association of Fire Fighters (IAFF), and the National Volunteer Fire Council (NVFC) to reduce the number of firefighters killed while responding to or returning from the emergency scene or while working at roadway emergency scenes. The USFA and the NVFC developed the Emergency Vehicle Safe Operations for Volunteer and Small Combination Emergency Service Organizations Program (Figure 1.3). This web-based educational program includes an emergency vehicle safety best practices self-assessment, standard operating guideline (SOG) examples, and behavioral motivation techniques to enhance emergency vehicle safety. This program also discusses critical safety issues of volunteer firefighting.

The USFA and IAFF developed a similar web and computer-based training and educational program—Improving Apparatus Response and Roadway Operations Safety in the Career Fire Service. This program discusses critical emergency vehicle safety issues such as seatbelt use, intersection safety, roadway operations safety on crowded interstates and local roads, and driver training. Instructor and participant guides and PowerPoint® slides are included.

In 2010, the IAFF also released a report entitled "Best Practices for Emergency Vehicle and Roadway Operations Safety in the Emergency Services" (Figure 1.4). This report was funded by the National Institute of Justice (NIJ), part of the U.S. Department of Justice (DOJ) and was produced under a cooperative agreement with the USFA. This report provides the latest information on all aspects of response and roadway scene for many of the disciplines who respond to emergency incidents including as police, fire, and emergency medical services (EMS) agencies.

The USFA and the IAFC developed "IAFC Policies & Procedures for Emergency Vehicle Safety" (Figure 1.5). This web-based document provides guidance for developing the basic policies and procedures required to support the safe and effective operation of all fire and emergency vehicles, including fire apparatus, rescue vehicles, ambulances, command and support units, POVs, and any other vehicles operated by fire department members in the performance of their duties. Links to each of these three programs are included in Appendix B: Resource Websites and Information Sources.

The original edition of this "Traffic Incident Management Systems" (TIMS) report was released in 2008 as part of a cooperative agreement between the UFSA and the International Fire Service Training Association (IFSTA) at Oklahoma State University (OSU). The project was funded by the DOT Federal Highway Administration (FHWA). This latest 2011 edition of TIMS was developed in response to the release of the 2009 edition of the DOT/FHWA's *Manual on Uniform Traffic Control Devices for Streets and Highways* (MUTCD). Changes in the 2009 MUTCD affected the content of the 2008 TIMS report and once again the DOT/FHWA funded the USFA to work with IFSTA to provide an updated report.

Figure 1.3. "Emergency Vehicle Safe Operations for Volunteer and Small Combination Emergency Service Organizations."

Figure 1.4. The IAFF released this publication in 2010.

Figure 1.5. "IAFC Policies & Procedures for Emergency Vehicle Safety."

IFSTA also completed a separate cooperative agreement with the USFA for the development of the "Emergency Vehicle Visibility and Conspicuity Study" (FA-323) that was released in August of 2009 (Figure 1.6). This report was also funded by the NIJ, part of the DOJ. This report provides detailed information on effective types of emergency lighting devices and retroreflective markings used on emergency vehicles. The report shows the connection between effective conspicuity and improved responder safety.

The USFA has also developed another resource related to response and roadway safety titled "Alive on Arrival." This two-page flyer provides tips for safe emergency vehicle operations. It focuses specifically on the roles of the apparatus operator and the passengers on board the apparatus. The complete document may be reviewed at: www.usfa.fema.gov/downloads/pdf/publications/fa_255f.pdf

Figure 1.6. The USFA worked with IFSTA to release the "Emergency Vehicle Visibility and Conspicuity Study" in 2009.

Other Government Initiatives for Roadway Safety

In addition to the various USFA-based programs dedicated to roadway response and roadway incident safety, there are numerous other programs at the Federal level that are having a major, positive impact on this issue. A few of these are described below.

Federal Highway Administration Traffic Incident Management Website

The FHWA Office of Operations operates an Emergency Transportation Operations (ETO) website featuring information on the ETO for disasters, Traffic Planning for Special Events (PSE), and Traffic Incident Management (TIM) programs. The FHWA, through the ETO programs, provides tools, guidance, capacity building and good practices that aid local and State DOTs and their partners in their efforts to improve transportation network efficiency and public/responder safety when a **nonrecurring** event either interrupts or overwhelms transportation operations. Nonrecurring events may range from traffic incidents to PSE to ETO for disasters. Work in ETO program areas focuses on using highway operational tools to enhance mobility and motorist and responder safety. Partnerships in ETO program areas involve nontraditional transportation stakeholders since ETO programs involve transportation, public safety (fire, rescue, EMS, law enforcement), and emergency management communities. ETO, as a discipline, spans a full range of activities from transportation-centric (fender benders) to those where transportation is a critical response component (e.g., hurricane evacuations). This web page can be viewed at: http://ops.fhwa.dot.gov/eto_tim_pse/about/tim.htm

Federal Highway Administration Traffic Incident Management Handbook

In 2010, the FHWA released a new edition of their *Traffic Incident Management Handbook*. This text includes the latest advances in TIM programs and practices across the United States. It also covers the latest innovations in TIM tools and technologies. This new edition supersedes the 2000 edition of the same title. It can be downloaded at no charge from: http://ops.fhwa.dot.gov/eto_tim_pse/publications/timhandbook/tim_handbook.pdf

National Traffic Incident Management Coalition

Launched in 2004, the National Traffic Incident Management Coalition (NTIMC) is a multidisciplinary partnership forum spanning the public safety and transportation communities to coordinate experiences, knowledge, practices, and ideas. NTIMC is committed to safer and more efficient management of all incidents that occur on, or substantially affect, the Nation's roadways in order to enhance the safety of onscene responders and of motorists passing or approaching a roadway incident; strengthen services to incident victims and to stranded motorists; and reduce incident delay and costs to the traveling public and commercial carriers.

One of the subjects that has been developed by the NTIMC is the "National Unified Goal for Traffic Incident Management: Working Together for Improved Safety, Clearance, and Communications." The goal of the NTIMC is to achieve three major objectives of the National Unified Goal (NUG) through 18 strategies. Key strategies include recommended practices for multidisciplinary TIM operations and communications, multidisciplinary TIM training, goals for performance and progress, promotion of beneficial, technologies, and partnerships to promote driver awareness. More information on the NTIMC can be found at: http://timcoalition.org/?siteid=41 Additional information on the NUG can be located at: www.transportation.org/sites/ntimc/docs/NUG%20Unified%20Goal-Nov07.pdf

Data Collection

Most agencies that collect and report data on firefighter injuries and deaths, such as the USFA and the National Fire Protection Association (NFPA), combine response-related casualties with roadway scene casualties into a single "vehicle-related" casualty reporting area. Of the two, clearly response-related injuries and deaths account for the majority of these casualties. This is why response-related issues were the primary focus of many of the previous USFA projects discussed earlier in this section.

When the two areas are analyzed separately, it becomes evident that injuries and deaths that occur at roadway emergency scenes are a major concern to emergency responders. The purpose of this report is to focus on the causes of firefighter injuries and deaths when working on roadway incidents. This report will focus on the causes of these incidents and provide strategies for mitigating them in the future. The occurrence and severity of these incidents can be reduced through proper roadway incident scene tactics and incident management, information which will be covered in the remaining chapters of this document.

The remainder of this chapter focuses upon statistics and causal information on these types of incidents. Although the remaining chapters of this report focus on roadway incident scene issues, some data on response-related injuries and deaths are also provided below to put the overall vehicle-related injury and death problem in perspective. In some cases, such as the topic of secondary collisions at roadway scene operations, the two are directly related.

Firefighter Fatalities

From 1996 to 2010, vehicle collisions claimed 253 firefighter lives and another 70 firefighters were lost as a result of being struck by a vehicle. Between 1996 and 2010, vehicle collisions/struck-by incidents accounted for 22 percent of all fatalities.

Table 1.1 provides a summary analysis of firefighter fatalities occurring in vehicle collisions and those struck by a vehicle while working on an emergency scene for the period from 1996 to 2010.

Table 1.1. Firefighter Fatalities in Vehicle Collisions and Struck by Vehicles (1996–2010)

Year	Total Deaths	Vehicle Collision	Struck by Vehicle	Percent of Total Deaths
1996	96	22	3	26
1997	94	14	5	20
1998	91	14	5	21
1999	112	11	6	15
2000	102	15	7	22
2001*	102	17	4	21
2002	100	20	6	26
2003	111	28	6	31
2004	110	19	7	24
2005	99	20	3	23
2006	92	11	5	17
2007	105	25	1	25
2008	107	18	4	21
2009	77	10	4	18
2010	72	9	4	18
Totals	**1,470**	**253**	**70**	**22**

*   The 2001 statistics do not include the 343 firefighters who perished as a result of the terrorist attack on New York City. The tragic loss of these firefighters was a statistical anomaly that would improperly skew the results of this issue.

Note: Total death figures from 2004 to 2010 do not include deaths that qualified solely under the Hometown Heroes Act of 2003.

Source: USFA, *Firefighter Fatalities in the United States* (1996–2010).

The types of vehicles involved in fatal collisions have remained consistent over this time period as well; POVs continue to be the most common vehicle involved in firefighter fatalities responding to and returning from an incident. Water tankers continue to be the most common fire apparatus involved in fatal collisions.

A report released by the Centers for Disease Control and Prevention (CDC) in 2010 also provides some interesting comparative data related to this study. The CDC report titled "Fatal Injuries Among Volunteer Workers—United States, 2003–2007" looked at the causes of deaths in all areas of volunteerism in the United States. This report noted that firefighters accounted for 109 deaths (38 percent) of the 287 fatal injuries to volunteers of all types. The report notes that 62 of the firefighter deaths were response and roadway incident scene related.

Perhaps most interesting to note in relation to the topic of this document was the fact that the CDC report showed that 53 percent of the total fatalities experienced by volunteers in all disciplines were vehicle related. This figure is very consistent with the fire service's own experience in this area. What this number may be telling us is that although any number of injuries and deaths is unacceptable, the number of vehicle-related deaths that the fire service experiences is not out of line with those in the general population of the United States. This does not mean, however, that we cannot improve upon those statistics.

Firefighter Injuries

Table 1.2 shows the summary of firefighter injuries occurring during response to and return from 1995 to 2010, the most recent years available at the time this report was written. What is statistically interesting in these numbers is the fact that while vehicle-related deaths account for a fairly significant percentage (second leading cause overall) of firefighter deaths, they actually account for only a small percentage of overall firefighter injuries.

Table 1.2. Firefighter Injuries Responding To/Returning From Incidents (1995–2010)

Year	Responding and Returning Injuries	Crash Injuries	Crash Injuries as a Percent of All Firefighter Injuries
1995	5,230	1,140	1.2
1996	5,315	1,150	1.3
1997	5,410	1,530	1.8
1998	7,070	1,365	1.6
1999	5,890	965	1.1
2000	4,700	1,160	1.4
2001	4,640	1,100	1.3
2002	5,805	1,250	1.5
2003	5,200	935	1.2
2004	4,840	1,200	1.6
2005	5,455	1,245	1.6
2006	4,745	1,460	1.8
2007	4,925	1,035	1.1
2008	4,965	740	0.9
2009	4,965	920	1.0
2010	4,380	850	1.0

Interestingly, these numbers tend to mirror the fire service's experience with cardiac-related injuries and deaths. Heart attacks and strokes are the leading killers of firefighters. On average, these events are responsible for 40–50 percent of firefighter deaths annually. However, cardiac events account for less than 2 percent of all firefighter injuries. What this tells us about both cardiac and vehicle-related events is that while they tend to be lower in frequency in the grand scheme of overall firefighter casualties, when they do occur they are serious events.

Secondary Collisions

A collision that occurs as a result of distraction or congestion from a prior incident is considered a secondary collision (Figure 1.7). There are many documented incidents resulting in both responder and civilian injuries or deaths as the result of secondary collisions. However, there is no specific database that allows for retrieval of the total numbers or any condition (e.g., weather, lighting, apparatus placement) related to the collisions. A DOT report indicated that approximately 18 percent of all traffic fatalities nationwide occur as a result of secondary collisions.

Figure 1.7. Secondary collisions account for 18 percent of all civilian traffic fatalities.

The Minnesota DOT references two studies in their Incident Management Program that estimate approximately 15 percent of all collisions result from an earlier incident. What must be remembered is that a secondary collision is often more serious than the original collision, especially if it occurs between free-flowing and stopped traffic. Secondary collision is an area where more studies and data are needed.

Law enforcement personnel are very cognizant of the likelihood and severity of secondary collisions. This often translates into one of the causes of friction that sometimes occurs between police officers and other emergency responders at the scene of roadway incidents. The police are under pressure to keeping traffic flowing and clear the scene as soon as possible, as this helps to minimize traffic delays and reduce the possibility of a secondary collision. In their view, the more apparatus and people brought to an incident, the more time it will take to eventually clear the scene, putting more sources of contact for secondary collisions on the roadway. The needs of both agencies must be balanced. This needs to be done in preincident planning and interagency cooperation. Trying to iron these issues out while standing in the roadway at an incident is rarely successful.

For the purpose of this report it must be realized that the majority of firefighter struck-by incidents fall into the category of secondary collisions. Most of the time, the only reason that firefighters are in the roadway in a position to be struck is because they are operating at an incident that already occurred. The principal purpose of much of the information contained in the remainder of this report is aimed at the prevention of secondary collisions.

Factors Influencing the Occurrence of Roadway Scene Incidents

Modern fire departments deliver a full range of fire, rescue, and EMS to handle virtually every type of emergency that may occur in a jurisdiction. These emergencies can happen at any time and in any location. Many of the emergencies that fire departments routinely respond to happen on the roadway. These include vehicle collisions, pedestrian collisions, vehicle fires, medical emergencies, and hazardous materials incidents. Other incidents may not actually occur on the roadway but require responders to deliver their services from the roadway, such as a medical emergency in a house next to the road.

Figure 1.8. All emergency responder should wear reflective vests when operating on the roadway.

In order to reduce the frequency of firefighters being struck by vehicles during the performance of their duties, it is important to understand some of the more common causes that lead to these incidents. The following is a summary of causal factors that have been noted in incident reports and through experience to be responsible for firefighters and other emergency responders coming in contact with other vehicles at a roadway incident scene.

- **Lack of training**—Responders are not trained on the hazards associated with roadway incidents and the proper ways to minimize these occurrences. They also may not be appropriately trained to work with other agencies.

- **Lack of situational awareness**—Responders fail to recognize the dangers associated with a particular roadway situation they are facing due to insufficient training or lack of experience.

- **Failure to establish a proper temporary traffic control (TTC) zone**—Many fire departments do not have sufficient training, equipment, or standard operating procedures (SOPs) for the correct way to set up a properly marked work area when operating at a roadway incident scene. Cases have also been noted where the responders did have good training, equipment, and SOPs, but for whatever reason failed to use or follow them (Figure 1.8).

- **Improper positioning of apparatus**—Numerous cases have been cited where apparatus was not positioned to the fullest advantage of the incident. In some cases the apparatus was not positioned in a manner that protected the work area. In other cases apparatus was unnecessarily positioned in the roadway.

- **Inappropriate use of scene lighting**—Inappropriate use of vehicle headlights, warning lights, and floodlights can confuse or blind approaching motorists (Figure 1.9). This causes them to strike an emergency vehicle, responder, or other vehicle in the incident area.

Figure 1.9. Scene light should not be blinding to oncoming motorists. *Courtesy of Ron Moore, McKinney, TX, Fire Department.*

- **Failure to use safety equipment**—Responders working in the roadway must wear appropriate protective garments and use all available traffic-control devices in order to prevent being struck by oncoming traffic.

- **Careless, inattentive, or impaired drivers**—Even when we try to do everything correctly, we must be cognizant of the fact that there are drivers out there who will not react correctly to the altered traffic pattern that occurs at a roadway incident. This may result in them driving into our workspace.

- **Reduced vision driving conditions**—Although firefighters may be struck by vehicles in virtually any condition, the chances of an incident occurring are greater during obscured vision conditions, including darkness, fog, rain, snow, and blinding sunshine.

- **Altered traffic patterns**—Drivers may be confused by the traffic control measures used at an incident scene or those being employed in a construction zone.

- **Lack of advanced warning devices**—Advanced warning signs and messages prepare the motorist for the conditions that he/she will soon encounter.

Other Considerations Relative to Roadway Incident Scenes

Fire service personnel need to look beyond the obvious, immediate concerns when considering the implications and impacts of roadway incident scenes. Taking a broader view of the subject will reveal some issues that fire service personnel and agencies should be more concerned about. It also gives keen insight into some of the major concerns held by other agencies with responsibility for roadway incident response. The fire service's failure to recognize these other concerns is one of the frequent sources of conflict that occurs between responding agencies at a roadway incident scene. Of course, the reverse is true as well.

Economic Impact

Some of the economic impacts of roadway incident scenes are quite obvious, while others may not be so apparent. Vehicle collisions have immediate and long-term economic effects on both the individual and society. Costs are both direct (those that are the result of the collision and resultant injury/fatality) and indirect (overall cost to society). These costs apply to both the victims of the primary incident and any responders who may be involved in a secondary incident and include, but are not necessarily limited to

- **Property damage**—Many of these costs are obvious and include the value of vehicles, cargo, roadways, negative impact to freight movement, adjacent property, and other items damaged in the incident.

- **Medical cost**—These costs include emergency room and inpatient costs, followup visits, physical therapy, rehabilitation, prescriptions, prosthetic devices, and home modifications for both the original victims of the incident and any responders who may be injured in a secondary collision.

- **Emergency services cost**—This includes the cost of providing police, EMS, and fire department response to the original incident and the additional costs of a secondary incident. In many cases the costs associated with providing service to the second incident will exceed those of the original incident.

- **Investigation cost**—The cost includes time spent investigating the incident and writing reports for primary and secondary incidents. In the case of fatal incidents these costs increase exponentially over injury or noninjury incidents.

- **Legal cost**—This includes fees, court costs, and overtime costs associated with civil litigation resulting from primary and secondary incidents.

- **Vocational rehabilitation**—This is the cost of job or career retraining required as a result of disability caused by roadway incident scene injuries.

- **Replacement employees**—Employers will often have to hire temporary help or pay other people overtime to cover the position of an injured employee.

- **Disability/Retirement income**—These costs occur when employees, including firefighters, cannot return to work.

- **Market productivity reduction**—This cost includes lost wages and benefits over the victim's remaining lifespan.

- **Insurance administration**—This is the administrative cost associated with processing insurance claims and attorney costs.

- **Travel delay**—This cost is the value of travel time delay for persons not involved in the collision, but who are delayed by the resulting traffic congestion. This is covered in more detail below.

- **Psychosocial impact**—This includes the cost of emotional trauma that inhibits, limits, or otherwise negatively influences a person's life.

- **Functional capacity**—This includes the long-term changes in a person's ability to function in daily living.

- **New operational costs**—This is the cost of developing new procedures and training to improve safety at future incidents.

Impact of Travel Delay Resulting From Vehicle Collisions

DOT and law enforcement officials try to minimize lane blockages not only because of fear of a secondary collision, but also because they realize the economic impact it has on those who become delayed in the resultant congestion. A general rule of thumb is that for every minute a lane of traffic is blocked by an incident, it results in 4 minutes of congestion. The FHWA estimates that the Nation loses 1.3 billion vehicle hours of delay due to incident congestion each year, at a cost of almost $10 billion. This does not take into consideration the cost of wasted fuel and environmental damage by idling vehicles in incident-related lanes of stopped traffic.

Every driver reacts differently to an unexpected incident. Reactions include slamming on the brakes, swerving into another lane, or just slowing down in order to gawk at the event. Regardless of the response, it creates a wave that progressively slows following traffic. Table 1.3 shows the reduction of vehicular traffic in relation to the location of the incident on a three-lane freeway (three lanes in each direction).

Table 1.3. Incident Effects of Blocking Lanes on Three-Lane Freeway

Incident Location	Capacity Reduction
Shoulder	17%
1 Lane Blocked	49%
2 Lanes Blocked	83%
3 Lanes Blocked	100%

Source: NHTSA, *Highway Safety Desk Book.*

It should be noted that the figures in Table 1.3 do not take into consideration the slowdowns that also typically take place in the opposing lanes of traffic due to curiosity, rubbernecking, and confusion caused by warning lights.

Several studies have been conducted to determine the cost of travel delay as the result of vehicle collisions. Lan and Hu's (2000) study in Minneapolis-St. Paul, MN, found an average of 5,057 hours of delay per heavy truck crash and 2,405 hours per crash without heavy vehicles involved. The study collected data on 289 heavy truck crashes and 3,762 other crashes.

NHTSA found that travel delay cost was $25.6 billion, or 11 percent of total collision costs, in 2000. Costs were calculated based only on police-reported crashes using the premise that any substantial impact on traffic would attract the attention of police. The costs per hour of delay were calculated using 60 percent of the wage rate for noncommercial drivers and 100 percent for commercial drivers. Table 1.4 shows a breakdown of the hours of delay by roadway type.

Table 1.4. Hours of Delay per Heavy Vehicle Crash by Roadway Type and Location (2000)

Roadway Type	Property Damage Only	Injury	Fatality
Urban			
Interstate	2,260	7,344	21,749
Other Freeway	1,766	5,737	16,990
Major Arterial	949	3,082	9,127
Minor Arterial	594	1,929	5,711
Collector	31	102	301
Local Street	9	28	83
Rural			
Interstate	814	2,646	7,835
Major Arterial	416	1,350	3,999
Minor Arterial	255	829	2,454
Major Collector	10	24	100
Minor Collector	4	14	42
Local Street	1	4	12

Note: Delay on local streets includes vehicles unable to exit from driveways. Each hour of delay in urban areas is valued at $13.86 and $16.49 in rural areas. Cost differential is due to differences in vehicle occupancy.
Source: NHSTA.

Project Goals
The USFA and the DOT/FHWA formed a partnership with IFSTA to research and identify effective technical guidance and training programs for fire and emergency service providers in TIMS. The initial version of this report was released in 2008.

The purpose of this project is to enhance responder safety and provide guidance to local-level fire departments on compliance with the 2009 edition of DOT's MUTCD and the National Incident Management System Consortium's (NIMSC's) *Model Procedures Guide for Incidents Involving Structural Fire Fighting, High Rise, Multi-Casualty, Highway, and Large-Scale Incidents Using NIMS-ICS* (also known as *IMS Book 1*). The information contained in this document should help enhance firefighter operational effectiveness, reduce potential liability, and enhance responder and motorist safety at roadway emergency scenes.

Figure 1.10. The MUTCD.

Manual of Uniform Traffic Control Devices for Streets and Highways
The effective use of approved traffic-control devices promotes highway safety and efficiency by providing for orderly movement of all road users. The MUTCD contains the basic principles that govern the design and use of traffic-control devices for all streets and highways, regardless of the public agency having jurisdiction (Figure 1.10). MUTCD requirements for TTC

devices are covered in Chapter 3 "Equipment to Improve Highway Safety" of this document. Chapter 6 "Best Practices and Other Sources of Information for Effective Highway Incident Operations" is particularly relevant to emergency highway operations, which are covered in depth in Chapter 4 "Setting Up Safe Traffic Incident Management Areas" of this document.

Experience shows that it is critical to integrate all-response agencies on highway incidents. The original "Model Procedures Guide for Highway Incidents" was developed by the NIMSC in cooperation with the DOT and it applies the organizational principles of Incident Management Systems (IMS) to generic highway incidents (Figure 1.11). It concentrated on the integration of all responders into a unified effort. The guide supported all response disciplines (fire, EMS, transportation, law enforcement, and public works) to address their specific tactical needs, while retaining the overall IMS structure. The information in this initial document was eventually absorbed into the *Model Procedures Guide for Incidents Involving Structural Fire Fighting, High Rise, Multi-Casualty, Highway, and Large-Scale Incidents Using NIMS-ICS* document. This information is reviewed in depth in Chapter 5 "Preincident Planning and Incident Command for Roadway Incidents" of this document.

Figure 1.11. "The Model Procedures Guide for Highway Incidents."

Chapter 2 Incident Case Studies

It is important to be aware of the numerical data and statistics on fatalities related to fire department response and roadway scene operations. This data gives a sense of the magnitude of the problem we are facing. However, it is also important to review several specific incidents in order to identify the factors involved and show the personal side of these tragedies. This chapter presents selected cases on firefighter pedestrian fatalities that were identified through the data obtained from the U.S. Fire Administration (USFA) firefighter fatalities studies over the past several years prior to the development of this report. As you read these cases studies, think about how many times you have been in a similar position or situation, yet did not fall victim to a collision.

Case Study 1

On October 27, 2003, at 2137 hours, volunteer members of a combination fire department responded to a report of a smoking generator at a road construction site. Seven volunteer firefighters in three apparatus responded. They determined that the problem was electrical in nature and notified the contractor who owned the equipment. One piece of apparatus, the brush truck, and three firefighters stayed on the scene to wait for the contractor to come and tend to the equipment.

When the contractor arrived, the assistant chief briefed the contractor on the situation and made preparations to leave the scene. As the brush-truck crew departed, they stopped at the entrance to the construction site to replace barricades they had moved upon entering. All three fire personnel got out of the brush truck, which was parked with its engine running and headlights and emergency lights on. As the assistant chief reached for a barricade, one of the firefighters noticed a white pickup truck headed towards them, fishtailing and apparently moving much faster than the posted 20-mile per hour (mph) speed limit. The firefighter yelled a warning to the other personnel and he and the other firefighter dove for cover. The pickup failed to make the sharp turn necessary to detour around the construction site. The pickup hit the left front corner of the brush truck. The vehicle then struck the assistant chief, who was standing toward the rear of the brush truck, and dragged him 60 feet before coming to a stop. The pickup lost its front left wheel in the crash and the assistant chief was partially pinned underneath the front of it, which was resting on the ground.

The other two firefighters ran back to the brush truck to call for assistance and get equipment. In the meantime, the driver of the pickup left the scene on foot. Medical care was provided by the firefighters on the scene and by responding paramedics. Despite their efforts, treatment was discontinued at the scene and the assistant chief was pronounced dead at 2348 hours.

Police searched through the night for the driver of the pickup involved in the crash but did not find him. The driver turned himself in the next day, admitting that his blood alcohol level was more than 0.10, the State's legal limit, at the time of the crash. He had been drinking at several bars before losing control of the pickup. After leaving the scene, he passed out in the yard of a nearby house and woke the next morning unaware of what had happened. The driver pled guilty to criminal vehicular homicide.

Causal Factors for the Incident

- The driver of the striking vehicle was reported to be under the influence of alcohol and was driving too fast for conditions.

- Visibility was decreased due to darkness at the time of the collision.

- The firefighters failed to exercise situational awareness and take appropriate precautions to prevent being struck by oncoming traffic.

Case Study 2

On January 9, 2001, at 1642 hours, a fire department was dispatched to a reported motor-vehicle crash with downed powerlines. At the time of the crash, the weather was reported as light snow with high winds causing limited visibility.

The department's assistant chief responded to the scene in his personal vehicle. Upon arrival, he reported that a vehicle had struck a power pole, and powerlines were down, but there were no injuries. He secured the scene and requested that the road be closed at the intersection of the State highway and a local road, 1.8 miles north of the crash site. The road was reported as having loose, wet snow with ice under the snow. Traffic was reported as unusually high due to a sporting event being held at a nearby school. There was a traffic signal at the intersection that was to be closed. The State highway had yellow caution lights and the cross street had stop signs and red lights.

Two firefighters proceeded to this intersection to close the road. There were no flares, cones, or signs posted on the roadway or at the intersection. The two firefighters were in street clothes, with no reflective vests, belts, or coats.

At 1720 hours, a civilian driver stopped in the intersection, signaling to make a left turn onto the closed road. One firefighter walked over to inform the driver that the road was closed due to the crash and downed powerlines. At 1722 hours, he stepped backward away from the driver's side window and a pickup truck traveling the other direction at approximately 20–25 mph struck him. The driver of the pickup reported applying the brakes as soon as he saw the firefighter step into his lane, however, the pickup slid on the slippery roadway and struck him.

The firefighter was thrown approximately 32 feet and pinned underneath a pickup in the opposing lane that was stopped in traffic. Ambulances responding to a simultaneous call were diverted to provide care for him. Firefighters and civilians at the scene lifted the pickup off of him by hand. He was first transported to a local hospital and then transferred by air ambulance to a regional trauma center. He was pronounced dead at 0323 hours on January 10 from a massive closed-head injury, pulmonary contusion, and chest injuries.

Additional information on this incident is available in National Institute of Occupational Safety and Health (NIOSH) "Fire Fighter Fatality Investigation and Prevention Program," report number 2001-07. The report is available for review at: www.cdc.gov/niosh/fire/reports/face200107.html

Causal Factors for the Incident

- Visibility was decreased due to darkness at the time of the collision.

- The firefighters failed to exercise situational awareness and take appropriate precautions to prevent being struck by oncoming traffic.

- The firefighters were not wearing appropriate retroreflective protective clothing.

- Road conditions at the time of the incident were poor (snow and ice) and may have prevented the striking vehicle from slowing or stopping in time to avoid the collision.

Case Study 3

On March 13, 2004, at 1654 hours, a fire department was dispatched to a vehicle and brush fire on a four-lane highway. The engine arrived at 1704 hours to find a fully involved minivan on the side of the road. Although a State highway patrol trooper was on the scene, the trooper had not slowed or diverted traffic and both northbound lanes were open. The engine was parked upwind of the burning vehicle.

As a firefighter stretched hose to begin fire suppression, the wind shifted and caused smoke to obscure visibility for oncoming motorists. The firefighter was struck by a Chevy Corsica that was driven through the smoke. He was thrown on top of the apparatus and then fell to the ground, where he died instantly.

The driver of the Corsica left the scene but was apprehended about an hour later. She proved to be a 28-year-old undocumented immigrant who was driving without a license. The driver told investigators that she thought that she had hit a cone, despite the fact that pieces of the firefighter's protective clothing were lodged in her windshield.

Subsequently, the driver pled no contest to leaving the scene of an accident and driving without a license and was sentenced to 2 years in prison.

In May of 2005, the County Training Officers Association adopted a standard set of procedures for highway incidents. The procedures include warning signs and high-visibility vests.

Causal Factors for the Incident

- Visibility was decreased due to smoke blowing across the roadway at the time of the collision.

- The firefighters failed to exercise situational awareness and take appropriate precautions to prevent being struck by oncoming traffic.

- The roadway was left open to traffic even though visibility was near zero because of smoke.

Case Study 4

On March 25, 2002, a truck company performed a required fire training exercise. The exercise involved search-and-rescue drills using machine-made smoke and mannequins.

The fire captain was working with other firefighters to recover and repack hose on the apparatus following the drill. During this process, a civilian vehicle entered the barricaded area where the firefighters were working at a high rate of speed, striking the captain and another firefighter.

The captain received serious injuries. The firefighter standing next to him received nonlife-threatening injuries. The captain was aggressively treated by firefighters and paramedics at the scene and transported to the hospital. His treatment continued upon arrival at the hospital, but he had suffered a massive head injury. Despite the efforts of responders and hospital staff, he died as a result of his injuries. The autopsy determined his death was due to skull fractures, subarachnoid hemorrhage, and cerebral edema. The police investigation classified the incident as vehicular manslaughter.

The driver of the car that struck him was arrested at the scene and later charged with driving while impaired by alcohol and prescription drugs.

Causal Factors for the Incident

- The driver of the striking vehicle was reported to be under the influence of alcohol and drugs at the time of the collision.

- Visibility was decreased due to darkness at the time of the collision.

Case Study 5

On December 23, 2003, at 0238 hours, a truck company was dispatched to assist an ambulance in responding to a vehicle crash on an expressway. Per the department's standard operating guidelines (SOGs), the truck was positioned to protect the ambulance and the crash scene from the flow of traffic, blocking the inside and center outbound lanes. State police also set flares to mark the scene.

Once it was determined that there were no injuries in the initial collision, the truck company was advised to return to service. The fire lieutenant was in the process of checking the truck to make sure all equipment had been replaced and that the compartment doors were closed. As the lieutenant was checking the exposed side of the apparatus, a 1997 Oldsmobile Cutlass illegally crossed over the center lane to cut in front of a tractor trailer in the outside lane, in an attempt to circumvent the crash scene. The Oldsmobile struck the tractor trailer on the front passenger side, causing it to spin counterclockwise and strike the lieutenant, pinning him between the car and the rear bumper of the fire truck. His legs were crushed by the impact and he died less than 12 hours later after suffering massive blood loss and kidney and heart failure.

The 26-year-old driver of the Oldsmobile had a blood alcohol level of .132 percent at the time of the crash, well above the State limit of .08. He also had a history of traffic violations in the State, where he never held a valid driver's license. He was charged with drunk driving and reckless homicide.

Causal Factors for the Incident

- The driver of the striking vehicle was reported to be under the influence of alcohol and was driving too fast for conditions.

- Visibility was decreased due to darkness at the time of the collision.

- The lieutenant failed to exercise situational awareness and take appropriate precautions to prevent being struck by oncoming traffic. He should not have placed himself between oncoming traffic and the exposed side of the apparatus.

Case Study 6

On July 1, 2002, at 0708 hours, a volunteer fire department and police personnel were dispatched to a vehicle fire on the right shoulder of the local interstate. A municipal police officer was first on the scene, parking his vehicle, with emergency lights activated, 30 feet behind the incident vehicle. An engine company with four firefighters on it arrived at 0712 hours and was positioned on the shoulder approximately 50 feet in front of the incident vehicle. The brush truck, with the captain on it, arrived at 0715 hours and was parked approximately 100 feet behind the incident vehicle, with all emergency lights activated. At the point where the incident occurred, the highway was straight and level and the road was dry.

The firefighters on the engine were working under the hood of the incident vehicle while the captain and the police officer stood near the passenger door talking with the driver. A northbound passenger car in the left lane was rear-ended by a pickup truck pulling a fifth-wheel camper. The passenger car skidded toward the shoulder, hit the police car, and then struck the captain, two other firefighters, the driver of the incident vehicle, and the police officer. It then impacted the incident vehicle, which was propelled approximately 50 feet and lodged under the rear of the engine company. The passenger vehicle came to rest about 50 feet behind the engine. The pickup crossed the median into the southbound traffic lane and then left the scene.

A State highway patrol officer witnessed the incident and radioed for assistance. The captain was found unresponsive, lying on the right shoulder of the highway just north of where the passenger vehicle came to a stop. After advanced life support (ALS) efforts, he was transported by air ambulance to a nearby hospital where he was later pronounced dead. The police officer was found unconscious, lying near the right rear tire of the passenger vehicle. Two firefighters and the owner of the incident vehicle were thrown onto the grassy area east of the right northbound shoulder. All three were injured, but conscious. Two firefighters jumped clear of the vehicles and escaped injury.

The captain was killed as the result of multiple traumatic injuries including a ruptured aorta. Additional information on this incident is available in NIOSH "Fire Fighter Fatality Investigation and Prevention Program," report number 2002-38. The report is available for review at: www.cdc.gov/niosh/fire/reports/face200238.html

Causal Factors for the Incident

- The firefighters failed to exercise situational awareness and take appropriate precautions to prevent being struck by oncoming traffic. The engine company was not parked in a manner that shielded the work area.

- Neither police nor fire personnel made any attempt to properly mark the incident scene or route traffic away from the work area.

Case Study 7

On the evening of February 3, 2004, a volunteer fire department was dispatched to a rollover motor-vehicle crash with injuries on a four-lane highway. The crash had blocked the right lane. A firefighter, wearing a reflective vest, was standing in that lane about 200 feet upstream of the crash scene to slow traffic and direct vehicles to move into the left lane. The firefighter was standing in the right-hand lane of two westbound lanes of the highway. He was wearing a retroreflective vest.

An automobile in the left lane passing the firefighter suddenly slowed and the vehicle behind it swerved to the right to avoid rear-ending it. In swerving, this vehicle entered the right-hand lane where the firefighter was standing. He was struck by the vehicle and thrown 136 feet into the ditch beside the road. He died of traumatic injuries on his way to the hospital.

His widow later sued the driver of the vehicle that struck him, a 19-year-old man who was not injured in the crash, as well as his parents, under a State law that allows parents of teenagers to be held responsible for the driving actions of their children. The driver of the car later pled guilty to careless driving involving a death.

Causal Factors for the Incident

- The driver of the striking vehicle was driving too fast for conditions.

- Visibility was decreased due to darkness at the time of the collision.

- The firefighters failed to exercise situational awareness and take appropriate precautions to prevent being struck by oncoming traffic.

- There were no signs or other traffic markers being used to direct traffic away from the flagger or incident scene.

Case Study 8

On March 19, 2003, at 0237 hours, a volunteer fire department was dispatched to a traffic incident with reported minor injuries in the eastbound lane of an interstate highway. The dispatcher realized that the incident was actually in a neighboring department's service area and she notified that department. That department dispatched an engine to the scene and requested mutual aid in the form of an ambulance because they were short on manpower.

A lieutenant from the first department that was dispatched started towards the fire station in his personal vehicle, but diverted straight to the scene. There was heavy fog at the station and the responding captain announced on the radio for all personnel to use caution. The lieutenant acknowledged the fog warning.

The first-arriving firefighter responded westbound and parked his vehicle in the median directly across from the incident, turning off his headlights but leaving his emergency flashers on. He notified the dispatcher that only one person had sustained hand injuries at the incident. The mutual-aid department uses the Incident Command System (ICS), but this firefighter did not take command because the incident was in the neighboring department's service area.

A paramedic had arrived before the first firefighter and parked her personal vehicle on the eastbound outside shoulder near the incident. A county sheriff's deputy was also on the scene. No traffic control had been established and all the vehicles involved in the original incident were parked on the shoulder or off the roadway.

At approximately 0259 hours, the lieutenant arrived and parked his personal vehicle behind the first-arriving firefighter's, in the median on the westbound side of the interstate. He exited his vehicle; he was wearing street clothes and did not put on his vest with reflective trim.

The driver of an eastbound tractor-trailer heard Citizens Band (CB) radio traffic regarding the incident, moved to the inside lane, and slowed to 48 to 50 mph. He saw the emergency lights on the vehicles parked on the eastbound outside shoulder and saw other nonemergency vehicles parked on the inside westbound shoulder.

As he passed the incident, the truck driver checked his right mirror to see if he had cleared the scene. When he looked back to the front, he saw the lieutenant step into the eastbound lane of traffic. He was unable to stop and struck him with the right front of the truck just to the left of the center divider line. He came to a controlled stop on the shoulder approximately 598 feet beyond the point of impact.

The lieutenant was thrown by the impact to the grassy median approximately 170 feet east of the point of impact. Others on the scene checked him, but he was obviously deceased. No charges were filed against the truck driver.

Additional information on this incident is available in NIOSH "Fire Fighter Fatality Investigation and Prevention Program," report number 2003-13. The report is available for review at: www.cdc.gov/niosh/fire/reports/face200313.html The State fire marshal also prepared a thorough report on this incident. That report is available at: www.tdi.state.tx.us/fire/fmloddinvesti.html

Causal Factors for the Incident

- Visibility was decreased due to darkness at the time of the collision.

- The lieutenant failed to exercise situational awareness and take appropriate precautions to prevent being struck by oncoming traffic.

- The lieutenant failed to don retroreflective personal protective equipment (PPE) that had been provided to him.

Case Study 9

On December 21, 2004, at 1645 hours, a fire department was dispatched to a vehicle crash. The first-arriving unit, a brush truck, found that the incident was actually on the border of the neighboring county. The incident was not technically a crash, as a vehicle had driven into a ditch at that location. Prior to the fire department's arrival, the people involved had been able to get their vehicle out of the ditch without assistance. The fire chief and one firefighter arrived on an engine shortly thereafter, only to learn from the first-arriving firefighter that they were not needed.

The two units proceeded south to find a driveway where they could turn around and return to the station. Although the chief asked him not to, the firefighter exited the engine to assist in allowing the truck to turn around. He took a traffic flashlight with him.

According to the traffic crash report, the driver of a white pickup truck headed northbound noticed the shadow of someone walking across the roadway and surmised the person was headed to his mailbox. As he approached the driveway, he saw the firefighter standing in his lane of traffic. He hit his brakes and swerved to the left, in an attempt to avoid hitting the firefighter, but the maneuver was not successful. The ambulance that had been dispatched to the original incident proceeded in and took the firefighter to the hospital, where he was pronounced dead. The cause of death was listed as multiple traumas.

Causal Factors for the Incident

- Visibility was decreased due to darkness at the time of the collision.

- The firefighter failed to exercise situational awareness and take appropriate precautions to prevent being struck by oncoming traffic.

- The firefighter failed to don retroreflective PPE while working in the roadway.

Case Study 10

On January 7, 2006, at 0715 hours, the fire department's shift had just come onduty when fire companies were dispatched to a series of weather-related motor-vehicle crashes in a curve on an eastbound interstate. According to a responding police officer, the road went from wet to black ice in a matter of minutes; there was no indication that ice had formed until vehicles began to slip and crash. A total of 13 vehicles were involved in 6 crashes along that stretch of roadway, supporting the finding that the road iced over very quickly and drivers had no knowledge of the need to slow down.

As the firefighter and her partner approached a pickup truck that had been involved in a crash to check on the occupants, another pickup came around the curve and lost control, striking a median wall and the first pickup. Someone yelled a warning to the firefighters, who began to move out of the way. Her partner was able to avoid the pickup, but she was struck by the front quarter panel of the vehicle on the driver's side. She was wearing her turnout gear and her helmet was knocked off by the impact. She was thrown an unknown distance to the east of the crash.

Firefighters coming to her aid found her lying on her side and unresponsive. She was quickly transported to the hospital and put on life support. However, she suffered closed-head trauma and was taken off of life support late in the afternoon of January 12, 2006. She died the next morning.

The district attorney's office declined to press charges against the driver of the pickup that hit her, citing the fact that no witnesses reported him driving in an erratic or unsafe manner. Blood tests done the day of the crash indicated that he was not under the influence of alcohol or drugs at the time.

Causal Factors for the Incident

- Visibility was decreased due to darkness at the time of the collision.

- Road conditions were poor due to ice on the roadway.

- The firefighters failed to exercise situational awareness and take appropriate precautions to prevent being struck by oncoming traffic.

- Neither police nor fire personnel made any attempt to close the roadway or shield the work area from approaching traffic.

Case Study 11

On June 29, 2001, at 2358 hours, a volunteer fire department was dispatched to a vehicle fire with a building exposure. Per departmental standard operating procedure (SOP), one properly-attired firefighter was to respond to the intersection near the fire station to assist with traffic control as the fire apparatus left the station. Because he lived close by, one fire police officer usually performed this role, as he did this evening.

The fire police officer was wearing reflective safety gear consisting of reflective safety helmet, high-visibility strobe light, high-visibility safety vest, and strobe traffic wand. The intersection was well lit by properly-operating mercury vapor street lights. Also per SOP, the responding fire engine came to a complete stop at the intersection. As it did, the lieutenant on the engine saw a pickup truck coming down the road faster than the posted 35-mph speed limit.

The fire police officer had his back to the pickup, but he turned and saw it coming toward him as the fire engine stopped. He took one step forward into the other lane to avoid the pickup, but it was over the double yellow lane marker and struck him. His body impacted the hood and cab of the pickup and was thrown approximately 75 feet forward, landing in a driveway.

The lieutenant radioed the dispatch center that a firefighter had been struck and the crew on the engine immediately went to his aid. A firefighter/emergency medical technician (EMT) who was responding to the original incident witnessed the event and provided care. The fire police officer was transported to the hospital, where he was pronounced dead.

The State police investigated his death and placed the driver of the pickup under arrest for driving while intoxicated and vehicular manslaughter.

Causal Factors for the Incident

- The driver of the striking vehicle was reported to be under the influence of alcohol and was driving too fast for conditions.

Case Study 12

On March 20, 2002, at approximately 1430 hours, a volunteer fire department was dispatched to a motor vehicle crash on the interstate in the southbound lane. There was a thunderstorm with heavy rains in progress in the area. One firefighter was the first volunteer to reach the scene, coming in from the north in his personal vehicle and crossing the median to park on the outbound shoulder in front of the original crash. He was wearing street clothes, jeans, and a light-colored shirt.

While he was on the scene, another crash occurred approximately 150 yards south of the first incident. He walked along the outside shoulder and approached the vehicle involved in the second crash on the passenger side to assess any injuries. He radioed the driver of the responding engine that there were no major injuries and that he could slow his response.

Approximately 2 minutes later, a motorist attempted to move from the right to the left lane but lost control of the vehicle. The automobile skidded off the road, traveled along the outer shoulder of the southbound lane, and struck the firefighter. The impact threw the firefighter into passing traffic, where he was hit by a pickup truck and thrown into the median. By that time, a sheriff's deputy had arrived on the scene and radioed for assistance. A rescue unit with two paramedics from a mutual-aid department self-dispatched to the scene and attended to the firefighter, who was unresponsive with no pulse or respirations. He was transported to the hospital, where he was pronounced dead.

Additional information on this incident is available in NIOSH "Fire Fighter Fatality Investigation and Prevention Program," report number 2002-13. The report is available for review at: www.cdc.gov/niosh/fire/reports/face200213.html

Causal Factors for the Incident

- Visibility was decreased due to heavy rain at the time of the collision.

- The firefighter failed to exercise situational awareness and take appropriate precautions to prevent being struck by oncoming traffic.

- The firefighter was not wearing retroreflective PPE.

- Road conditions were poor at the time of the incident due to heavy rainfall.

Case Study 13

On July 27, 2007, a fire department responded to a tractor-trailer fire on an interstate highway. Three fire trucks were on the scene and were parked on the right shoulder and the first traffic lane to the left of the shoulder. Safety cones had been placed in the roadway and all apparatus warning lights were activated.

The response of the State police was significantly delayed. The Incident Commander (IC) declined offers of assistance from local law enforcement agencies offering traffic-control assistance, citing the lack of traffic on the highway.

At approximately 0415 hours, one firefighter was replacing equipment that had been used into a compartment on the driver's side of the vehicle. The firefighter was struck by a passing bus and thrown over 200 feet to the side of the road. The firefighter suffered fatal injuries. The bus driver was charged with negligent homicide and reckless driving.

Causal Factors for the Incident

- The chief officer refused the assistance of local law enforcement officials to assist with scene control. Never turn down qualified assistance in these situations.

- Personnel were operating between the apparatus and the oncoming flow of traffic.

Case Study 14

On June 14, 2008, a Sheriff's deputy and a volunteer assistant fire chief were fatally injured after being struck by a tractor-trailer on a four-lane highway at the scene of a previous motor-vehicle crash. Visibility at the time of the incident was described as near-zero due to fog and smoke from a fire on a nearby military range.

The truck driver attempted to slow his tractor-trailer down after encountering the smoke and fog and swerved suddenly to miss a vehicle parked on the highway. The tractor-trailer struck Sheriff's deputy #2's patrol car, positioned partially on the shoulder and left lane, in the right rear quarter panel. The patrol car skidded to the left, striking Sheriff's deputy #2 and knocking him into the median and injuring him. The tractor-trailer continued north in the left lane, striking the fire officer and Sheriff's deputy #1, killing them on impact. It is believed that Sheriff deputy #1 had just finished providing instructions to move the parked vehicle in the right northbound lane. The tractor-trailer then swerved right, striking a vehicle in the right northbound lane that was involved in the first northbound incident. The tractor-trailer finally came to rest against the rear doors of an ambulance parked in the left northbound lane.

The Highway Patrol estimated the speed of the tractor-trailer was 55 mph when approaching this area and 50 mph upon striking the first vehicle. The tractor-trailer tire skid marks before striking the patrol car were 54 feet in length and the tractor-trailer traveled 167 feet after striking the patrol car.

Causal Factors for the Incident

- There was a significant delay between the first responders arriving on the scene and addressing the visibility and roadway safety issues that were present.

- Emergency vehicles were not parked in a manner to protect the work areas of this roadway incident scene.

- Police and fire personnel must not operate outside of the safety zone that a properly-positioned emergency vehicle would have created.

- Traffic was allowed to continue through the incident scene despite extremely low-visibility conditions. In these situations, the approaching traffic, even at appropriate slow speeds, may not be able to see personnel or vehicles in time to prevent from striking them.

Summary

A review of these case studies finds that some of the factors that led to these deaths are within the control of firefighters and some are not. Of those that are within our control, we see multiple examples of basic safety procedures not being followed. In these case studies, we see firefighters who consistently do not recognize the danger signs present at the roadway scene, firefighters who fail to wear appropriate protective clothing, and fire and police agencies that do not take effective actions in guarding the incident scene and work area from oncoming traffic. The remainder of this manual is focused on information to improve the performance of firefighters and other emergency responders in these situations.

Chapter 3 Equipment to Improve Highway Safety

For several years now, the U.S. Department of Transportation (DOT) has been engaged in a program entitled Intelligent Transportation Systems (ITS). The goal of ITS is to improve transportation safety and mobility and enhance productivity through the use of advanced communications technologies. There are nine major initiatives within the ITS program. They include

1. Vehicle Infrastructure Integration (VII).

2. Next Generation 9-1-1.

3. Cooperative Intersection Collision Avoidance Systems.

4. Integrated Vehicle-Based Safety Systems.

5. Integrated Corridor Management Systems.

6. Clarus, the Nationwide Surface Transportation Weather Observing and Forecasting System.

7. Emergency Transportation Operations (ETO).

8. Mobility Services for All Americans.

9. Electronic Freight Management.

Much work has been done within the ETO section of ITS relative to the safety of firefighters and other first responders who are working on the roadway. One of the concepts being studied within this area of the project is the concept of using Traffic Incident Management Systems (TIMS) to reduce the effects of incident-related traffic congestion by decreasing the time necessary to detect incidents, the time for responding vehicles to arrive, and the time required for traffic to return to normal conditions. TIMS contributes to increasing emergency responders' safety at an incident scene both directly and indirectly.

Though many of the findings and features of the overall ITS project are not directly related to issues firefighters will work with or even be concerned about, they have a direct positive impact on the safety of firefighters who work on the roadway. The first portion of this chapter discusses some of these projects. This information is based on the Federal Highway Administration's (FHWA's) "Intelligent Transportation Systems Benefits and Costs: 2003 Update" found at: www.itsdocs.fhwa.dot.gov/jpodocs/repts_te/13772.html#2.4

Much of what firefighters need to know about traffic control and safe operations on the highway is contained in a DOT document titled Manual on Uniform Traffic Control Devices for Streets and Highways (MUTCD). The current version of the MUTCD available at the time of this report was the 2009 edition. Each State had 2 years to review and adopt the most current edition of the MUTCD "as is" or make changes that make its version more stringent than the Federal version; State versions cannot be less stringent than the Federal version. The MUTCD, in essence, is the bible of roadway operations for all highway operations, both routine and emergency in nature. The MUTCD refers to the incidents emergency responders work on as traffic incident management areas (TIMAs) and states that effective temporary traffic control (TTC) measures must be in place at these scenes. The MUTCD states that the three primary functions of TTC at a TIMA are to

1. Move road users reasonably safely and expeditiously past or around the traffic incident.

2. Reduce the likelihood of secondary traffic collisions.

3. Preclude unnecessary use of the surrounding local road system.

The latter portions of this chapter discuss the appropriate types of MUTCD-compliant equipment that can be used to establish TTC at roadway emergency incidents. It also discusses some new technologies and equipment used outside the United States. The goal of this chapter is to provide firefighters with information on the correct types of equipment they should be using when working on the roadway.

Intelligent Transportation Systems Technologies to Improve Roadway Safety

This section details a selection of new technologies that the ITS program has advanced for the improvement of roadway safety and incident-scene safety. Many of these are only indirectly related to the role of firefighters, but firefighters should be aware of their existence and impact on the jobs they perform.

Traffic Surveillance Technology

The ITS program has been responsible for the development and installation of a wide variety of traffic surveillance and detection technologies, such as acoustic roadway detectors, still photos, and video camera systems (Figure 3.1). These technologies monitor traffic flow, detect deviations in traffic patterns, feed information to a traffic-control center, and notify responders of traffic conditions on the way or the best route to approach the scene. In some cases, the traffic control center is able to send emergency help before civilians on the scene are able to dial 9-1-1. Video-based systems may also be used to provide emergency responders with important information on the incident while they are still en route.

Figure 3.1. Traffic status and control can be monitored at high-tech command centers.

ITS is also responsible for advances in wireless enhanced 9-1-1 systems and automatic collision notification (ACN) systems. Although not directly involved with emergency response, these technologies can help identify the problem early, contact the appropriate help, and divert traffic through announcements to the public. This helps to reduce congestion, speed response to the scene, and prevent secondary collisions.

Mayday and Automatic Collision Notification Systems

ACN systems can impact both firefighter and motorist safety. These systems transmit voice and data to an emergency call center upon manual activation when the driver presses a button (Mayday) or automatically when they are triggered by onboard safety equipment such as airbag deployment. The OnStar© system that is used in General Motors (GM) vehicles is perhaps the most recognizable of these systems. These units use in-vehicle crash sensors, Global Positioning System (GPS) technology, and wireless communications to supply call centers with crash location and, in some cases, the number of injured passengers and nature of injuries. Advanced ACN products can assist in determining the type of equipment needed (basic life support (BLS) or advanced life support (ALS)), mode of transport (air or ground), and location of the nearest trauma center. Although anecdotal, reports suggest that ACN systems positively impact victim outcomes by reducing time to emergency medical care.

The National Highway Traffic Safety Administration's (NHTSA's) 1998 Strategic Plan noted that 24 percent of all fatal crashes in the United States occur on rural roads. However, this relatively small percentage of crashes accounts for nearly 59 percent of all crash deaths. One factor that contributed to the disproportionately high-fatality rate for rural crash victims was a delay in delivering emergency medical services (EMS) to the scene. Included in these deaths are many volunteer firefighter deaths responding to rural incidents or fire stations in private vehicles. The highest percentage of deaths in actual fire apparatus crashes are those involving fire department tankers (tenders), which also tend to occur in rural areas. ACN systems could help decrease those fatalities by lessening the response time of emergency medical care for those involved in the collision.

Two studies on ACN are worth reviewing. Under a grant from DOT, Harris County, TX, installed ACN systems in 500 police and fire department vehicles in 2002 for a 2-year pilot study. The ACN fed information directly to a roadside assistance provider in Boston, MA, who called the vehicle to confirm the crash and verify the identities of occupants. They then forwarded the information to a Colorado-based telecommunications and public safety technology provider, who used the vehicle's location to route the data to the appropriate public safety answering point (PSAP) using existing 9-1-1 systems. Simultaneously, the occupants' demographic data was forwarded to a Virginia-based technology provider who generated an injury prediction algorithm and related that to the trauma center. An enhanced version of this system with real-time reporting was tested in 2008.

The Minnesota DOT, in partnership with the Mayo Clinic (Rochester, MN), launched a lower-tech system in 2003. In the Minnesota system, telematics system providers (TSPs) relay emergency calls and caller location obtained from the GPS unit in the vehicles equipped with an ACN system to the PSAP on 9-1-1 priority voice communications lines. The TSP also transmits additional data on to the Condition Acquisition and Reporting System secure data network. Responding agencies are able to access the incident data from the Internet, according to their data access privilege classifications.

ACN systems are becoming more common as standard installation on new cars, and there are also aftermarket products. The cost of ACN devices range from approximately $400 to $1,900. These units appear to hold a great deal of promise in improving incident reporting and, thus, emergency response. However, there is a fee for the service of the TSPs, ranging from $10 to $27 per month, so if the fee has not been paid, the ACN will be inactive.

Freeway Service Patrols

ITS has encouraged the development and operation of freeway service patrols. Freeway service patrols operate in many major metropolitan areas, as well as some suburban and rural areas. These are often State DOT programs and consist of a fleet of light-duty trucks that have two-way radio communication with a traffic control center and are usually equipped with motorist assist supplies, traffic cones, a lighted vehicle arrow board, and, in some cases, extendable floodlights (Figure 3.2).

Figure 3.2. Many State transportation departments operate roadway safety units. *Courtesy of the Georgia DOT Heroes Program.*

While the primary focus of these units is to monitor roadway conditions and provide assistance to disabled motorists, these patrol vehicles are also typically dispatched to roadway incidents to assist other emergency responders with traffic control. Depending on local protocols, dispatch of these units may be automatic, by request of the Incident Commander (IC), or from law enforcement personnel. State DOT representatives should be included as part of the traffic Incident Management Team (IMT) to identify criteria and standard operating procedures (SOPs) for incorporating DOT resources into roadway-scene responses to aid in traffic control and reduce incident-related delay.

Figure 3.3. CMS can warn drivers of impending hazards.

Changeable Message Signs

Changeable message signs (CMS) are becoming more common on the Nation's freeways (Figure 3.3). They provide a versatile means of communicating information to drivers and can be invaluable in alerting oncoming traffic to an emergency incident. Although, in some locations, incident management personnel can directly post incident-related information to CMS, usually messages are posted by transportation management center personnel. Emergency responders should be familiar with the procedures for contacting the agency that controls sign messages within their jurisdiction.

For CMS to be useful, the message must be concise and clear. CMS used on roadways with speed limits of 55 miles per hour (mph) or higher should be visible from one-half mile under both day and night conditions. The message should be designed to be legible from a minimum distance of 600 feet for nighttime conditions and 800 feet for normal daylight conditions. When environmental conditions that reduce visibility and legibility are present, or when the legibility distances stated in the previous sentences in this paragraph cannot be practically achieved, messages composed of fewer units of information should be used and consideration should be given to limiting the message to a single phase.

Each message shall consist of no more than two phases. Each phase shall consist of no more than three lines of text. The minimum time that an individual phase is displayed should be based on 1 second per word or 2 seconds per unit of information, whichever produces a lesser value. The display time for a phase should never be less than 2 seconds. The maximum cycle time of a two-phase message should be 8 seconds.

Messages should be concise, clear, and provide relevant information. All messages are printed in capital letters. The average driver traveling at a high rate of speed can handle eight-word messages of four to eight characters per word at 2 to 4 seconds per message. The message should consist of at least the problem and action and may contain an effect. For example, let's say the problem is an accident two miles ahead in the right lane. Drivers should expect delays and merge left. A two-pane CMS might read:

Panel 1: **ACCIDENT AHEAD TWO MILES**

Panel 2: **MERGE LEFT EXPECT DELAYS**

A one-panel might read: **ACCIDENT TWO MILES MERGE LEFT**

Portable CMS should be visible from one-half mile under both day and night conditions. Letter height should be a minimum of 18 inches and legible from at least 650 feet if the sign is mounted on a trailer or large truck. If mounted on service patrol trucks, letter height should be a minimum of 10 inches and visible from at least 330 feet (Figures 3.4a and 3.4b).

Figure 3.4a. Some portable message signs are mounted on a ground stand. *Courtesy of Ron Moore, McKinney, TX, Fire Department.*

Figure 3.4b. This portable message sign is attached to the rear of a fire department command vehicle. *Courtesy of Janet Wilmeth.*

Temporary Traffic Control Zones

Before getting into a detailed discussion of the types of equipment most emergency responders will use to assist with traffic control at a roadway emergency scene, it is first necessary to review the basic components of a TTC zone. The procedures for establishing these zones will be covered in more detail in Chapter 4 of this document. The MUTCD defines a TTC zone as:

"A TTC zone is an area of a highway where road user conditions are changed because of a <u>work zone, or an incident zone, or a planned special event</u> through the use of TTC devices, uniformed law enforcement officers, or other authorized personnel.

"A work zone is an area of a highway with construction, maintenance, or utility work activities. A work zone is typically marked by signs, channelizing devices, barriers, pavement markings, and/or work vehicles. It extends from the first warning sign or high-intensity rotating, flashing, oscillating, or strobe lights on a vehicle to the END ROAD WORK sign or the last TTC device.

"An incident <u>zone</u> is an area of a highway where temporary traffic controls are imposed by authorized officials in response to a traffic incident. It extends from the first warning device (such as a sign, light, or cone) to the last TTC device or to a point where road users return to the original lane alignment and are clear of the incident.

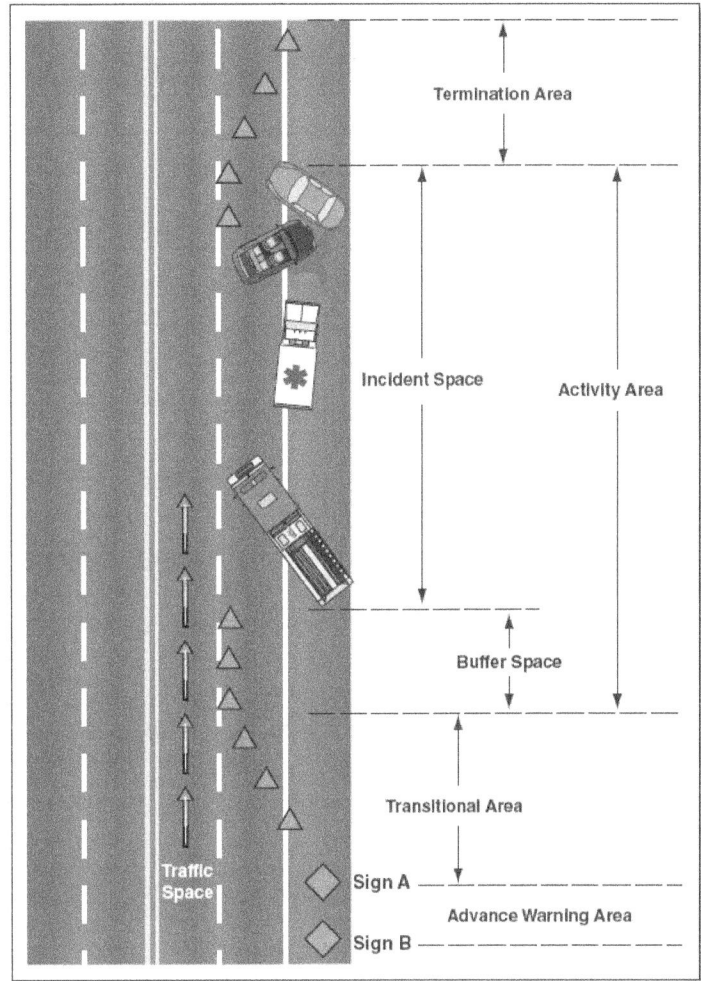

Figure 3.5. The typical parts of a TIMA.

"A planned special event often creates the need to establish altered traffic patterns to handle the increased traffic volumes generated by the event. The size of the TTC zone associated with a planned special event can be small, such as closing a street for a festival, or can extend throughout a municipality for larger events. The duration of the TTC zone is determined by the duration of the planned special event."

TTC zones may be established for a variety of reasons, including road maintenance, weather conditions, disabled vehicles, planned events, and emergency incidents. The MUTCD refers to emergency scenes on the roadway as traffic incident management areas (TIMAs). To be specific, emergency responders need to be familiar with the MUTCD procedures for establishing TTC at TIMAs.

Most TTC/TIMAs are divided into four areas (Figure 3.5). The **advanced warning area** is the section of highway where drivers are informed of the upcoming incident area. Because drivers on freeways are assuming uninterrupted traffic flow, the advance warning sign should be placed further back from the incident scene than on two-lane roads or urban streets. Table 3.1 shows the stopping sight distance as a function of speed.

Table 3.1. Stopping Sight Distance as a Function of Speed

Speed (mph)	Distance (ft)
20	115
25	155
30	200
35	250
40	305
45	360
50	425
55	495
60	570
65	645
70	730
75	820

The **transition area** is the section of the TTC zone where drivers are redirected from their normal path. This usually involves the creation of tapers using channelizing devices. Tapers may be used in both the transition and termination areas. The MUTCD designates the distance of cone placement to form the tapers based on the speed limit multiplied by the width of the lanes being closed off. This can be shown mathematically as follows:

TL = (LW x the # of lanes) x PSL

Where: TL = Taper length in ft

LW = Lane width in ft

PSL = Posted Speed Limit in mph

For example, suppose you are closing two lanes of an interstate highway whose speed limit is 75 mph. The lanes are 12-feet wide. In this example, the taper length would be calculated as follows:

TL = (LW x the # of lanes) x PSL

TL = (12 ft x 2 lanes) x 75 mph

TL = (24)(75)

TL = 1,800 ft

The **activity area** is the section of highway where the work activity or incident takes place. It is made up of the workspace, the traffic space, and the buffer space. The workspace is where the actual work activity occurs. The traffic space is the portion of the roadway used to route traffic through the incident area. The buffer space is the lateral and/or longitudinal area that separates traffic flow from the work area. The buffer space may provide some recovery space for an errant vehicle. The MUTCD (Section 6C.06) specifically states that "Neither work activity nor storage of equipment, vehicles, or material should occur within a buffer space."

The **termination area** is used to return drivers to their normal path. It ends at the last TTC device. Conditions and safety considerations may dictate the need for a longitudinal buffer space between the work area and the start of the downstream taper.

See Chapter 4 "Setting Up Safe Traffic Incident Management Areas" of this publication for more information on establishing TTC zones.

Channelizing Devices

Channelizing devices are used to warn drivers of conditions created by incident activities in or near the roadway and to guide drivers around the incident. Channelizing devices used during an emergency incident can include signs, cones, tubular markers, flares, directional arrows, and flaggers.

Signs

The MUTCD establishes specific color requirements for signs that will be used in different situations. The MUTCD (Section 6I.01) states that the required colors for warning signs used for TTC in TIMAs is fluorescent pink with black letters and border. In emergency situations where fluorescent pink signs are not available, older style signs with yellow backgrounds may be used (Section 6F.16). However, it is recommended that as fire departments and other emergency response agencies replace old signs or purchase new signs, the new signs be of the pink with black letter type.

Figure 3.6. "Emergency Scene Ahead" signs should be placed well ahead of an incident scene.

The MUTCD gives minimum direction on the required sizes for TTC signage. Where roadway or road-user conditions require greater emphasis, larger than standard size warning signs should be used, with the symbol or legend enlarged approximately in proportion to the outside dimensions of the overall sign. Departments with limited resources are advised to acquire larger signs, such as 48x48 inches, as they are suitable for most any situation. When a series of two or more advance warning signs is used, the closest sign to the TTC zone should be approximately 100 feet for low-speed urban streets to 1,000 feet or more for freeways and expressways (Section 6F.17). Exact distances are detailed in Chapter 4.

National Fire Protection Association (NFPA) 1500, *Standard on Fire Department Occupational Safety and Health Program*, which applies to fire service agencies, also requires that a retroreflective fluorescent pink highway safety sign be deployed as advance warning any time a fire department vehicle is used in a blocking mode at a highway incident. NFPA requires the wording "EMERGENCY SCENE AHEAD" for the sign (Figure 3.6). In essence, this mirrors the MUTCD requirement.

Cones

Traffic cones are perhaps the most commonly used channelizing devices. Cones must be predominantly orange and made of a material that can be struck without causing damage to the impacting vehicle. Cones should be weighted enough that they will not be blown over or displaced by wind or moving traffic. It is important to understand that MUTCD (Section 6F.64) requirements for traffic cones used during the day and on low-speed roadways (≤ 40 mph) are different than for cones used at night and/or on freeway or high-speed roadways (≥ 45 mph).

For daytime and low-speed roadways, cones shall be not less than 18 inches in height. When used on freeways and other high-speed highways or at night on all highways, cones shall be a minimum of 28 inches in height. For nighttime use, cones shall be retroreflectorized or equipped with lighting devices

Figure 3.7. Traffic cones are useful in setting up a safe work zone. *Courtesy of Ron Moore, McKinney, TX, Fire Department.*

Figure 3.8. Traffic cones can be stored in a variety of places on an emergency vehicle.

Figure 3.9. Some traffic cones are equipped with flashing lights.

for maximum visibility. Retroreflectorization of cones that are 28 to 36 inches in height shall be provided by a 6-inch wide white band located 3 to 4 inches from the top of the cone and an additional 4-inch wide white band located approximately 2 inches below the 6-inch band (Figure 3.7).

Retroreflectorization of cones that are more than 36 inches in height shall be provided by horizontal, circumferential, alternating orange and white retroreflective stripes that are 4 to 6 inches wide. Each cone shall have a minimum of two orange and two white stripes with the top stripe being orange. Any nonretroreflective spaces between the orange and white stripes shall not exceed 3 inches in width.

The MUTCD does not specify whether the cones need to be of the solid or collapsible styles. Many fire departments choose to equip fire apparatus with collapsible cones, as they reduce the amount of required storage space. Others find unique, easily accessible locations to carry cones on the apparatus (Figure 3.8). There are a variety of options that can be used to increase the effectiveness of the cones, particularly in low-light situations. Cones are available that illuminate from within or are equipped with light strips that encircle them. Cones may also be equipped with flashers attached to the tops (Figure 3.9).

Flares

There are three basic types of flare devices that may be used in TTC zones. These devices include

1. Incendiary flares.

2. Chemical light sticks.

3. Light emitting diode (LED) flares.

Each of these devices is detailed in the following section. Additional information on all three types of these flares can be found in a U.S. Department of Justice (DOJ) report titled "Evaluation of Chemical and Electric Flares" at: www.ncjrs.gov/pdffiles1/nij/grants/224277.pdf

Incendiary Flares

Some form of incendiary road flare has been used to alert drivers to dangerous conditions for almost 100 years (Figure 3.10). Incendiary flares are self-sustaining. There are no concerns about battery life or corroding electrical parts. Incendiary flares burn at approximately 70 candela. By comparison, chemiluminescent light sticks are approximately 10 candela and a typical flashlight is 5 candela.

There are several concerns with the use of incendiary flares. Incendiary flares are classified as a flammable solid and must be stored according to specific guidelines. The chemicals in

Figure 3.10. Traffic flares create a bright pattern of light. *Courtesy of Ron Moore, McKinney, TX, Fire Department.*

standard incendiary road flares (strontium nitrate, potassium perchlorate, and sulfur with a sawdust/oil binder) are hazardous substances. Exposure to the chemicals causes corrosive injury to the eyes and irritation to the skin and respiratory tract. Lit flares can cause skin burns and destroy clothing and vehicle tires. Incendiary flares cannot be used at scenes with fuel spills, hazardous materials, high-fire risk conditions, or during high-wind conditions. Cleanup is often required after use. Emergency personnel must ensure that all flares that may pose a continuing ignition source or traffic hazard are removed from the scene before responders depart.

Figure 3.11a. These battery-operated LED flares can be placed on the roadway.

Chemical Light Sticks

Chemical light sticks generate chemiluminescence in an enclosed container, making them suitable for use in hazardous environments. Two different types of chemicals (usually luminal and oxalate) are stored within two tubes, an outer one and an inner glass vial. These two tubes are stored in a transparent plastic container. The glass vial floats in the outer tube's chemical. When the outer tube is bent or broken and shaken, the chemicals combine and start to glow. Glow time is between 6 to 12 hours.

Chemical light sticks are inexpensive and easy to store and use. However, once they are activated, they cannot be reused. LED light sticks are a reusable alternative to chemical light sticks. They are battery operated and will last about 20 hours if left on continuously; longer if turned on and off intermittently. Light sticks are not as bright as incendiary fusees or LED flares.

Figure 3.11b. These LED and chemical light stick flares are placed on a traffic cone.

Light Emitting Diode Flares

These devices use LEDs to project an extremely bright light, visible 360° from great distances. Depending on the manufacturer, the lights may be adjusted between a steady, flashing, or rotating mode. One manufacturer has a mode that emulates the flicker of an incendiary flare. The rotating and flashing signals put out by these units are nonhypnotic and nondisorienting. These units come in a variety of configurations; some lie flat on the ground, some can sit on stands, and some come with a bracket that attaches them to the top of a traffic cone (Figures 3.11a and 3.11b). Most use disposable or rechargeable AA or AAA batteries.

These units average approximately 90 to 100 hours of running time. They are sturdy, standing up to the weight of vehicle traffic and weatherproof. As of this writing, the cost of the units varied from $10 to $50 per unit.

Directional Arrow Boards

An arrow board is a sign with a matrix of elements capable of either flashing or sequential displays. Directional arrow boards can provide additional warning and directional information for merging and controlling drivers through/around a TTC zone. Directional arrow boards must be used in conjunction with other TTC devices such as channelizing equipment. There are four types of arrow boards. Type A is used on low-speed urban streets. Type B is used on intermediate-speed roadways and for maintenance or mobile operations on high-speed roadways. Type C is used in areas of high-speed, high-volume motor vehicle traffic. Type D is used on State or local authority authorized vehicles. Type A, B, and C arrow panels shall be a solid rectangle. Type D shall conform to the shape of the arrow. All arrow panel boards should be finished in nonreflective black. The minimum mounting height of an arrow is 7 feet from the roadway to the bottom of the board, except on vehicle-mounted board.

Figure 3.12a. This directional arrow is located on a traffic-control vehicle. *Courtesy of Jack Sullivan.*

Figure 3.12b. This flashing arrow board is designed into the rear of the apparatus.

It is becoming increasingly common for fire departments to mount directional arrow boards on apparatus (Figures 3.12a and 3.12b). When contemplating this, it is important to review the MUTCD requirements in Section 6F.61 to assure the arrow boards are compliant. Arrow boards should be capable of operating in all three modes:

1.   Flashing arrow, sequential arrow, or sequential chevron.

2.   Flashing double arrow.

3.   Flashing caution or alternating diamond mode.

Figure 3.13 shows these modes. The board must be capable of at least a 50 percent dimming from full brilliance for use during nighttime operation in order to not adversely affect oncoming driver vision. The size of the arrow on apparatus-mounted boards must be 48-inches long, and the width of the arrowhead must be 24 inches and must be visible at a minimum of one-half miles. It should be noted that many of the arrow boards and directional light bars currently located on apparatus do not meet this standard. If the arrowhead is not obvious to approaching traffic, it simply becomes another blinking yellow light. Although there is no specified height, vehicle-mounted arrow boards should be as high as practical, have remote controls, and the vehicle must have high-intensity rotating, flashing, oscillating, or strobe lights.

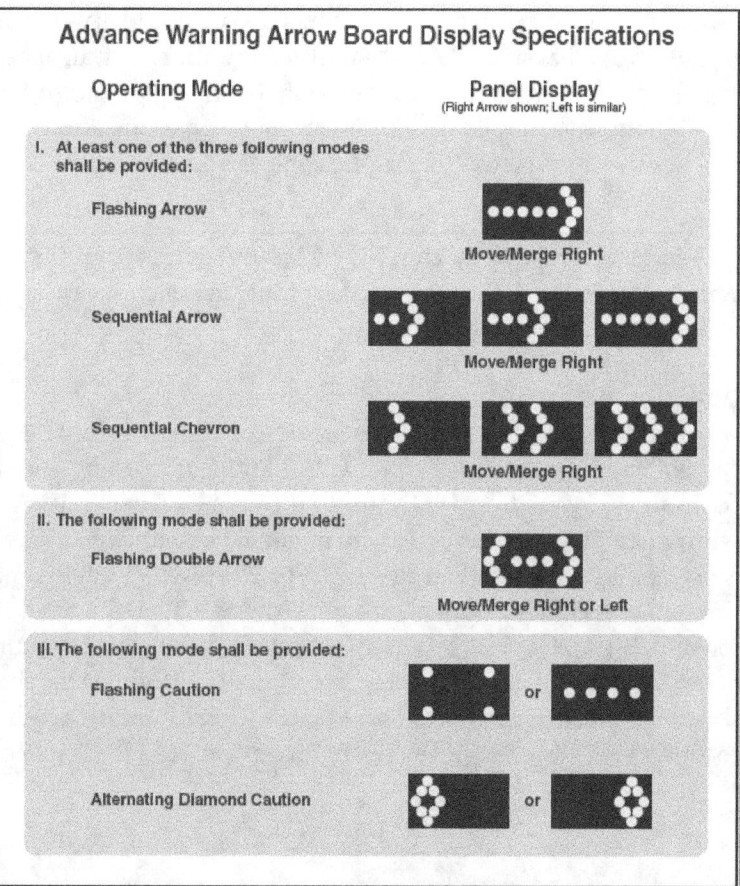

Figure 3.13. This illustration shows the various flashing arrow modes.

Barricades

Collisions involving multiple vehicles, collisions resulting in fatalities, or hazardous material spills may require a road closure. As part of an overall incident management plan, this type of incident would most likely involve the State DOT for incidents on major roadways and local or county street departments on surface roads. The freeway patrol units discussed earlier would be able to provide initial traffic control if available. Neither fire apparatus nor the freeway patrol units normally carry barricades. Thus, DOT resources would need to be dispatched to place barricades and other appropriate portable signs and TTC devices.

The MUTCD, Section 6-F, identifies four types of barricades (Figure 3.14). Rail-stripe width on all barricades 36 inches or over must be 6 inches. For barricades that are less than 36-inches wide, the rail stripe may be 4 inches. The side of the barricade facing traffic must have retroreflective rail faces. Warning lights on barricades are optional.

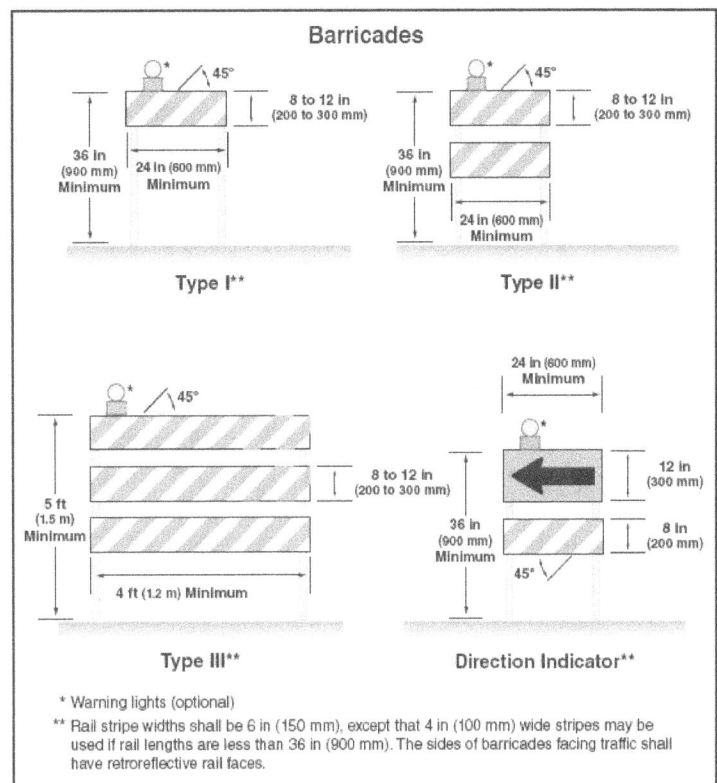

Figure 3.14. Types of traffic barricades.

Flagger Control

In many situations, it will be necessary to use emergency personnel to assist in directing traffic at a roadway incident, especially early in the incident. This section details the requirements for personnel who are assigned this function.

A flagger is a person who manually provides TTC. According to the MUTCD, Section 6E, a flagger is responsible for the safety of both emergency workers and the motoring public. Any person, including an emergency responder, who is assigned to direct traffic is considered a flagger and therefore must be trained and meet the MUTCD flagger requirements.

Many volunteer departments on the east coast of the United States use fire police to direct traffic at incident scenes. Fire police are members of the fire department who focus on providing roadway-scene safety protection functions and crowd control at incidents. This includes directing traffic, setting up signs and other blocking equipment, and securing incident scenes. In some jurisdictions, they are formally sworn in by the municipality as reserve police officers. In some cases, they operate apparatus specially equipped for these functions (Figure 3.15). In other

Figure 3.15. Some jurisdictions operate fire police vehicles to assist with traffic control. *Courtesy of Jack Sullivan.*

jurisdictions, firefighters may be assigned this function. Regardless of who is assigned this function, it is important to review the MUTCD qualifications for flaggers. Flaggers should have the following abilities:

- receive and communicate specific instructions;

- move and maneuver quickly;

- control signaling devices to provide clear and positive guidance to drivers;

- understand and apply safe traffic control practices; and

- recognize dangerous traffic situations and warn workers in sufficient time to avoid injury.

For daytime and nighttime activity, flaggers shall wear high-visibility safety apparel that meets the Performance Class 2 or 3 requirements of the American National Standards Institute (ANSI)/International Safety Equipment Association (ISEA) 107–2004 publication entitled "American National Standard for High-Visibility Apparel and Headwear" (see Section 1A.11) and labeled as meeting the ANSI 107-2004 standard performance for Class 2 or 3 risk exposure. The apparel background (outer) material color shall be fluorescent orange-red, fluorescent yellow-green, or a combination of the two as defined in the ANSI standard. The retroreflective material shall be orange, yellow, white, silver, yellow-green, or a fluorescent version of these colors, and shall be visible at a minimum distance of 1,000 feet. The retroreflective safety apparel shall be designed to clearly identify the wearer as a person.

In lieu of ANSI/ISEA 107-2004 apparel, public safety (law, fire, EMS) personnel within the TTC zone may wear high-visibility safety apparel that meets the performance requirements of the ANSI/ISEA 207-2006 publication entitled "American National Standard for High-Visibility Public Safety Vests" and labeled as ANSI 207-2006.

Departments that require personnel to perform flagger duties should ensure that those personnel complete a MUTCD-compliant flagger course. Fire officials may wish to consult local transportation officials for information on these courses within their jurisdiction.

Hand-Signaling Devices

Hand-signaling devices, such as STOP/SLOW paddles, flashlights/wands, and red flags, are used by flaggers to control drivers. The STOP/SLOW paddle (Figure 3.16) is the MUTCD-preferred hand-signaling device because it provides more positive guidance for drivers. The paddle is octagonal on a rigid handle. It must be at least 18-inches wide with letters at least 6-inches high. The background of the STOP side must be red with white letters and border, while the SLOW side must be orange with black letters and border. When used at night, the paddle must be retroreflectorized (MUTCD, Section 6E.03).

Flagger Location

Flaggers must be located so that approaching drivers have sufficient distance to stop at the intended stopping point or slow to merge lanes. Refer back to Table 3.1 to review the stopping sight distance as a function of speed and thus determine the

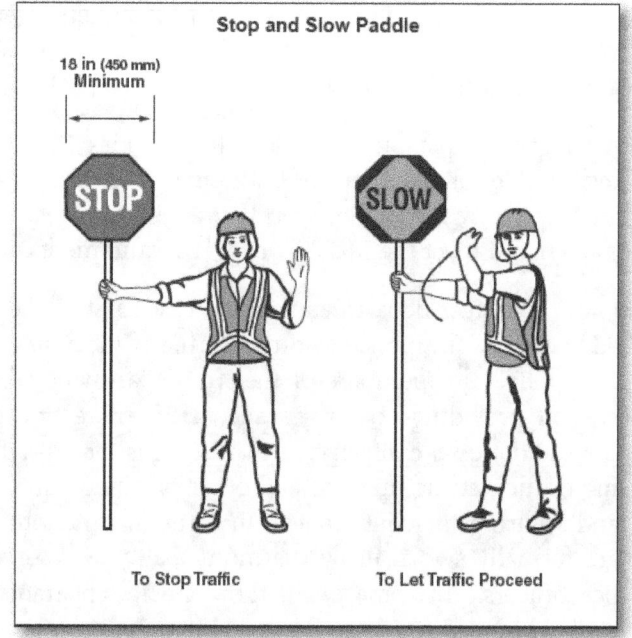

Figure 3.16. A lighted traffic paddle.

Figure 3.17 – A proper position for a flagger in the closed lane of traffic.

flagger location. The flagger should be far enough in advance of workers to warn them of approaching danger by out-of-control vehicles. The flagger should wear proper protective equipment as described below and always stand alone.

Flaggers should stand either on the shoulder adjacent to the lane being controlled or in the closed lane prior to stopping drivers (Figure 3.17). The flagger should only stand in the lane being used by moving traffic after traffic has been halted. The flagger should be clearly visible to the first-approaching driver at all times, as well as being visible to other drivers. Flaggers at emergency incidents must use extreme vigilance since there may not be an advance warning sign before traffic reaches the flagger. The use of hand movements alone without a paddle, flag, or other approved devices to control road users shall be prohibited except for law enforcement personnel or emergency responders at incident scenes.

The following three methods of signaling with paddles shall be used:

1. To stop road users, the flagger shall face road users and aim the STOP paddle face toward road users in a stationary position with the arm extended horizontally away from the body. The free arm shall be held with the palm of the hand above shoulder level toward approaching traffic.

2. To direct stopped road users to proceed, the flagger shall face road users with the SLOW paddle face aimed toward road users in a stationary position with the arm extended horizontally away from the body. The flagger shall motion with the free hand for road users to proceed.

3. To alert or slow traffic, the flagger shall face road users with the SLOW paddle face aimed toward road users in a stationary position with the arm extended horizontally away from the body.

Audible Warning Signals

The MUTCD suggests equipping flaggers with a horn or whistle to provide an audible warning to workers of oncoming danger. An air horn or compressed-gas horn would work well. If a whistle is used, make sure the necklace has a breakaway attachment allowing it to pull loose if caught on an object or moving vehicle.

The device used to warn workers of dangers when working at a traffic incident should be loud enough to be heard above the noise of traffic and any equipment being used by emergency workers. Ron Moore of "Firehouse Magazine" states that relying on a radio call may not be sufficient for all to hear during highway operations. The radio channel may be busy, not everyone on the scene may have a radio, or not everyone may be on the same channel.

High-Visibility Safety Apparel

Every year traffic increases, leading to more congestion and greater risk to emergency response personnel. Conditions at dawn, dusk, night, and during inclement weather increase the risk. Personnel visibility is imperative to responder safety. **Note:** Although all firefighter turnout clothing includes the use of retroreflective markings per the requirements of NFPA 1971, *Standard on Protective Ensembles for Structural Fire Fighting and Proximity Fire Fighting*, these requirements fall well short of meeting MUTCD requirements for safety garments to be worn on the roadway. Firefighters must wear additional protective garments when working on roadway emergency scenes. With the exception of DOT workers, most other responders to roadway incidents have normal clothing with no reflective markings whatsoever.

The importance of wearing retroreflective safety apparel is also cited in the 2008 "Effects of Warning Lamp Color and Intensity on Driver Vision" that was written by the Society of Automotive Engineers (SAE) with assistance from the U.S. Fire Administration (USFA) and DOJ. From this project, it was discovered that:

- Detection distances for pedestrians and emergency responders operating on the roadway at night wearing typical clothing are very short; shorter than typical required stopping distances.

- In contrast, detection distances for pedestrians and emergency responders operating on the roadway at night with retroreflective markings are very good.

There are two other documents with which firefighters and other public safety responders should be familiar. The first is ANSI/ISEA *American National Standard for High-Visibility Apparel* (ANSI 107). This document set the requirements for high-visibility safety apparel worn by public safety personnel (and all other highway workers) for many years and much of the equipment in use today was designed to this document. In 2007, ANSI/ISEA released a new standard, ANSI/ISEA 207-2006, *American National Standard for High-Visibility Public Safety Vests*. This document has more specific requirements for safety apparel that should be worn by firefighters and other public safety personnel who work on the highway.

American National Standards Institute/International Safety Equipment Association 107

The MUTCD specifies safety apparel

"meet the requirements of the American National Standards Institute/International Safety Equipment Association (ANSI/ISEA) *American National Standard for High-Visibility Apparel* and it must be labeled as meeting the ANSI 107-1999 standard performance for Class II risk exposure. The apparel background (outer) material color shall be either fluorescent orange-red or fluorescent yellow-green as defined in the standard. The retroreflective material shall be orange, yellow, white, silver, yellow-green, or a fluorescent version of these colors, and shall be visible at a minimum

Figure 3.18. ANSI Class II and Class III compliant garments. *Courtesy of Ron Moore, McKinney, TX, Fire Department.*

distance of 1,000 feet. The retro-reflective safety apparel shall be designed to clearly identify the wearer as a person. (This is particularly important for emergency workers among the flashing lights and other apparatus markings at the scene.) For nighttime activity, Class III risk exposure should be considered for flaggers" (Section 6E.02). Figure 3.18 shows ANSI Class II and Class III compliant garments.

After 5 years, ANSI/ISEA revised this standard and released ANSI/ISEA 107-2004. The new standard sets performance criteria and guidelines for the selection, design, and wearing of high-visibility safety clothing. It defines three protective classes based on background material, retroreflective material, and design and usage requirements. It also provides criteria to assist in determining the appropriate garment based on roadway hazards, work tasks, complexity of the work environment, and vehicular traffic and speed. Table 3.2 summarizes the classes.

Table 3.2. American National Standards Institute/International Safety Equipment Association Garment Classifications

Class	Intended Use	Worker Example
I	Activities that permit the wearer's full and undivided attention to approaching traffic. There should be ample separation of the worker from traffic, which should be traveling no faster than 25 mph.	• Parking lot attendants • Warehouse worker • Roadside "right of way" or sidewalk maintenance workers
II	Activities where greater visibility is necessary during inclement weather conditions or in work environments with risks that exceed those for Class I. Garments in this class also cover workers who perform tasks that divert their attention from approaching traffic or are in close proximity to passing vehicles traveling at 25 mph or higher.	• Forestry operations • Roadway construction, utility, and railway workers • School crossing guards • Delivery vehicle drivers • Emergency response and law enforcement personnel
III	Activities of workers who face serious hazards and often have high task loads that require attention away from their surroundings. Garments should provide enhanced visibility to more of the body, such as the arms and legs.	• Roadway construction personnel and flaggers • Utility workers • Survey crews • Emergency response personnel

Fabric

ANSI/ISEA 107-2004 specifies that the fabric must be tightly knit or woven for background coverage. Therefore, open mesh fabrics are not in compliance since they do not provide the background coverage or brightness to meet the standard. The fabric must also be stain and water-repellent. The standard also requires retesting the chromaticity (brightness and purity of color) of fabrics after a laboratory light-exposure test.

Fluorescence

Fluorescent fabrics absorb ultraviolet (UV) light of a certain wavelength and regenerate it into lower energy and longer wavelengths. This property makes the garments especially bright on cloudy days and at dawn and dusk, when UV light waves are high. Fluorescent fabric does not glow in the dark. The new standard requires certification of the fluorescent background fabric to specific chromaticity minimums. Although several colors are available, the most popular safety colors are lime/yellow and orange.

A 1990 survey conducted by the Minnesota DOT displayed four mannequins in fluorescent jumpsuits. Minnesota State Fair attendees were asked to choose the most visible mannequin. Fluorescent yellow was clearly the most visible color (Table 3.3). In addition, of 119 color-impaired attendees surveyed, 97 percent selected fluorescent yellow.

Table 3.3. Minnesota Fluorescent Color Survey, 1990

Color	Number of Participants
Yellow	5,796
Green	2,706
Orange	2,231
Pink	2,017

Retroreflectivity

Retroreflective fabrics are necessary to extend the same level of protection at night that fluorescent fabrics provide during daylight. Retroreflective fabric works like a mirror, reflecting light back to its source. The standard identifies the requirement (photometric performance) of retroreflective material alone or combined with fluorescent fabric. Photometric performance is measured by candle power (cd/lux/m²). There are two classes of retroreflectivity. Apparel must provide 360° of visibility, so the retroreflective striping must basically encircle the torso. All retroreflectors deteriorate with time. The rates of deterioration depend on the type of material, use, and exposure to the environment. Table 3.4 provides a summary of the ANSI/ISEA 107-2004 garment class requirement.

Table 3.4. 107-2004 American National Standards Institute/International Safety Equipment Association Garment Class Requirement

Requirement	Class I	Class II	Class III
Background material minimum area	217 in² (0.14 m²)	775 in² (0.5 m²)	1,240 in² (0.80 m²)
Retroreflective or combined-performance material used with background material	155 in² (0.10 m²)	201 in² (0.13 m²)	310 in² (0.20 m²)
Minimum width of retroreflective bands	310 in² (0.20 m²)	N/A	N/A
Minimum number of yds per retroreflective band width	1-inch (25 mm) or 2-inch (50 mm) combined-performance material (without background material)	1.378 in² (35 mm)	2 inch (50 mm)
Minimum number of yds per retroreflective band width	4.3 yds of 1-inch- (25 mm) wide bands 3.1 yds of 1.378-inch- (35 mm) wide bands 2.15 yds of 2-inch- (50 mm) wide bands	4 yds of 1.378-in²- (35 mm) wide bands 2.8 yds of 2-inch- (50 mm) wide bands	4.3 yds of 2-inch- (50 mm) wide bands

American National Standards Institute/International Safety Equipment Association 207

The revised ANSI/ISEA 107-2004 standard clearly prohibits any kind of sleeveless garment to be labeled Class III when worn alone. This change would have a significant effect on some emergency response departments. Because of these problems, a number of public safety organizations, led by the Emergency Responder Safety Institute (ERSI), lobbied DOT, ANSI, and ISEA for a specific standard for a vest to be used in the public safety sector.

A significant event related to the safety of emergency workers operating on the roadway occurred on December 8, 2006, with the release of ANSI/ISEA 207-2006. ANSI/ISEA 207-2006 establishes design, performance specifications, and use criteria for highly-visible vests that are used by public safety industries. The standard includes basic requirements such as vest dimensions, color, and materials performance, and also incorporates criteria for special features for users in fire, emergency medical, and law enforcement services.

These special features include easier access to belt-mounted equipment (guns for police, EMS tools, etc.) and the ability for vests to tear away from the body if they are caught on a moving vehicle. Vests labeled as ANSI 207-compliant should have breakaway features on the two shoulder seams, two sides, and in the front for a total of five breakaway points (Figure 3.19). The ERSI urges buyers to specify five breakaway points and accept no less than four breakaway points (all except the front closure) when ordering the public safety vests. ANSI/ISEA 207-2006 also allows for color-specific markings on the vest panel or trim to clearly and visibly distinguish between police, fire, and EMS responders. These colors include red for fire officials, blue for law enforcement, green for emergency responders, and orange for DOT officials.

When comparing the new ANSI/ISEA 207-2006 public safety vest standard to the ANSI/ISEA 107-2004 standard, the following distinctions should be noted:

Figure 3.19. An ANSI 207-compliant vest.

- ANSI/ISEA 207-2006 is for public safety responders only and is not intended to replace or be interchangeable with ANSI 107-2004 Class II requirements. In fact, the 450 in^2 of reflective material required of an ANSI 207 vest falls between the requirements for ANSI 107 Class I and II.

- Law enforcement officers performing traffic-control duties are still encouraged to follow ANSI 107 Class II or Class III guidelines whenever possible.

- A lesser background area requirement on ANSI/ISEA 207-2006 allows for short designs, giving tactical access to equipment belts.

- Retroreflective area requirements for ANSI/ISEA 207-2006 are the same as those for ANSI 107-2004, Class II vests.

- The new standard suggests use of many design options, such as breakaways, colored identifiers, loops, pockets, badge holders, and identification (ID) panels.

In addition to the ANSI/ISEA draft for public safety vests, the FHWA has released a "Notice of Proposed Rulemaking on Worker Visibility." This rule suggests that all workers (including emergency responders) on U.S. Federal-aid highways must wear high-visibility garments while performing their duties. The proposed rule references Class II and Class III garments, but not public safety vests. The ERSI submitted a comment to reference public safety vests in addition to Class II and III garments.

The standard will only affect law enforcement, emergency responders, fire officials, and DOT personnel sectors. It will improve the safety in multiagency incidents by improving visibility and identification. It will reduce confusion and enhance communication between agencies. Basic vest requirements will include

- vest dimensions;

- material performance;

- special design features for users in fire, emergency medical, and law enforcement services; and

- higher visibility (checkered color-coded reflective trim).

Fire Apparatus Safety Equipment

Many fire apparatus markings and devices are used to improve the safety of personnel riding in the apparatus, working at an incident, and working in close proximity to the apparatus. Most of these features are also addressed in the USFA publication "Emergency Vehicle Safety Initiative" (FA-272, 2004). This section will review those that have the most impact on safety at highway operations.

Restraints

Managing an incident scene appropriately is contingent on personnel arriving safely. Fatalities occurring as a result of apparatus collisions almost doubled in 2003–2004 over fatalities occurring in apparatus collisions from 1994 to 2002. The average number of fatalities from 1994 to 2002 was 12, compared to 22 for 2003–2004. This trend has continued to date. Lack of restraint-use continues to be a problem. Historically, only one in five firefighter fatalities in vehicle collisions were wearing restraints.

Figure 3.20. Red seatbelts making it easier for the officer to determine if the members are properly belted.

In addition, firefighters may often simply be sitting in vehicles that are parked for the purpose of blocking at roadway emergency scenes. If the fire apparatus is struck by a vehicle approaching the incident scene, unsecured firefighters could sustain serious injuries or be killed in that collision.

Figure 3.21. This pumper has the minimum NFPA 1901-compliant striping.

NFPA 1500 specifies the mandatory use of restraints during any response, whether emergency or nonemergency in nature. NFPA 1901, *Standard for Automotive Fire Apparatus*, requires red seatbelt webbing, making it easier to check compliance (Figure 3.20).

Figure 3.22. This apparatus has reflective markings and a stop sign on the inside of the cab doors.

Vehicle Striping

Previous editions of NFPA 1901 required a simple 4-inch wide retroreflective stripe that extends at least 50 percent of the length of the vehicle on each side and 25 percent of the width of the front of the vehicle (Figure 3.21). A graphic design that meets these parameters is an acceptable substitute. NFPA 1901 also requires retroreflective striping inside cab doors to maintain conspicuity and alert passing drivers to an open door (Figure 3.22). A major addition to the 2009 version of NFPA 1901 was the requirement for a European-style retroreflective chevron pattern to cover at least 50 percent of the rear-facing surface of the vehicle. The stripes must slope downward and away from the centerline of the vehicle at an angle of 45°. (Figure 3.23). Each stripe must be 6-inches wide and in an alternating pattern of red and yellow, fluorescent yellow, or fluorescent yellow-green.

All law enforcement and fire service agencies should make sure that all of their vehicles have lighting systems and reflective markings that are within the bounds of their State motor vehicle code and any other standards that apply. Departments that are unsure whether or not they are in compliance with the State motor vehicle code should seek assistance from their State police agency or DOT. Fire apparatus manufacturers are typically well-versed in the requirements of NFPA 1901, as well as NFPA 1906, *Standard for Wildland Fire Apparatus*, which sets similar design requirements for wildland fire apparatus. At the time this document was being developed, the NFPA was in the process of developing a new standard (NFPA 1917) for the design and construction of ambulances that is slated for release in late 2013.

Figure 3.23. Newer fire apparatus are required to have chevron markings on the rear of the vehicle.

In 2009, the International Fire Service Training Association (IFSTA) completed a cooperative research agreement project for the USFA and the DOJ's National Institute of Justice (NIJ) titled *Emergency Vehicle Visibility and Conspicuity Study*. The purpose of this study was to determine effective colors, patterns, and overall usage of these markings. The report looks at the European cross hatching and other reflective markings that are being used on emergency vehicles to determine the best choices for emergency vehicles. This report is meant to complement a different project that the USFA did with the SAE on the effectiveness of emergency vehicle warning lights.

Warning Lights

NFPA 1901 requires all fire apparatus have a system of optical warning lights in the upper and lower zones and on all four sides. The standard identifies two modes of emergency lighting. The "calling for right of way" is the light pattern used while the apparatus is in motion. The "blocking right of way" mode is the light pattern used while the apparatus is parked at the incident. The "best" light color(s) continue to be debated. A 1978 study by the National Institute of Standards and Technology (NIST) showed that as the number of flashing lights increases, the ability of drivers to quickly respond decreases. Strong stimuli holds central gaze and drivers tend to steer in the direction of gaze.

The MUTCD also addresses the use of warning lights at roadway incident scenes in Section 6I.05. The use of emergency lighting is essential, especially in the initial stages of a traffic incident. However, it only provides warning; it does not provide traffic control. Emergency lighting is often confusing to drivers, especially at night. Drivers approaching the incident from the opposite direction on a divided roadway are often distracted by the lights and slow their response, resulting in a hazard to themselves and others traveling in their direction. (It also often results in traffic congestion in the unaffected opposite lane(s) and increases the chance of a secondary collision.)

Emergency-vehicle lighting can be reduced if good traffic control has been established. If good traffic control is established through placement of advanced warning signs and TTC devices, responders can perform their tasks with minimal emergency-vehicle lighting. This is especially true at major incidents that involve a number of emergency vehicles. Departments should review policies on emergency-vehicle lighting, especially after a traffic incident scene is secured, with the aim of reducing the use of vehicle lighting as much as possible while not endangering those at the scene. Special consideration should be given to reducing or extinguishing forward-facing vehicle lighting, especially on divided roadways.

An internal lighting study conducted by the Phoenix Fire Department following a 1994 fatality suggested that a reduced level of amber (yellow) lighting was less likely to blind drivers and less likely to draw the interest and attention of passing drivers. As a result, the process began to reconfigure engines for all nonamber warning lights (clear, red, and blue) to go off when the apparatus parking brake was engaged. Amber lights on all four sides of the apparatus are the only functioning lights in the "blocking right of way" mode. Many other fire departments in the United States have also adopted this practice.

In 2007, the USFA, in partnership with the DOJ's NIJ, entered into a cooperative agreement with the SAE to look at the issue of nonblinding emergency-vehicle lighting. The SAE worked with the researchers at the University of Michigan Transportation Research Institute (UMTRI) to conduct this research. The results were published in a USFA report titled "Effects of Warning Light Color and Intensity on Driver Vision" in October 2008 (Figure 3.24).

Figure 3.24. This report was published by the USFA and the SAE.

This report was part of a program of research on how warning lights affect driver vision and how those lights can be designed to provide the most benefit for the safety of emergency vehicle operations. In order to understand the overall effects of warning lights on safety, it is necessary to know about the positive (intended) effects of the lights on vehicle conspicuity, as well as any negative (unintended) effects that the lights may have on factors such as glare and driver distraction. The report also provides information about how the colors and intensities of warning lights influence both positive and negative effects of such lights, in both daytime and nighttime lighting conditions. Color and intensity have received considerable attention in standards covering warning lights at the local, State, and national levels. Interest in these variables has recently increased because of the new options provided by the growing use of LED sources in warning lamps.

Participants in this study were selected to be reasonably representative of the driving public. Two groups, based on age, were chosen to ensure that some estimate could be made of how warning-light effects might change with driver age. A static field setting was used to simulate the most important visual circumstances of situations in which drivers respond to warning lights in actual traffic. Two vehicles with experimental warning lights were placed so that they would appear 90° apart in a simulated traffic scene as viewed by an experimental participant who was seated in a third vehicle. The four most commonly-used colors of warning lights in the emergency services were used (white, yellow, red, blue) and all four colors were presented at two levels of intensity. All intensity levels were high relative to current minimum requirements, since the greatest interest was in measuring potential benefits of high-intensity lamps in the day and possible problems with high-intensity lamps at night. Participants performed three tasks, under both day and night conditions:

1. Lamp search, in which the participant had to indicate as quickly as possible whether a flashing lamp was present on the right or left simulated emergency vehicle. This task was designed to capture the kind of visual performance that would be important when a driver tries to locate an emergency vehicle approaching an intersection on one of two possible paths. Faster performance for a certain type of lamp can be taken to mean that the lamp provides better conspicuity.

2. Pedestrian-responder search, in which the participant had to indicate as quickly as possible whether a pedestrian-responder wearing turnout gear was present near the right or left simulated emergency vehicle. This was designed to capture negative effects of the warning lamps on seeing pedestrian responders near an emergency vehicle. Slower performance for a certain type of lamp can be taken to mean that the lamp causes more interference with driver vision (e.g., glare or distraction).

3.   Numerical rating of the subjective conspicuity of warning lamps. This task was designed to provide a subjective measure of the visual effects of lamps, which may or may not show the same effects of color and intensity that are provided by the objective search tasks.

The results of all three tasks showed major differences between day and night conditions. Search for lights was easier during the night and search for pedestrians was easier during the day. The large differences in performance between night and day add support, and some level of quantification, to the idea that the most significant improvements that can be made in warning lights may be in adopting different light levels for night and day.

Over the range of light intensity that was used, there were improvements with higher intensity for the light-search task during the day, but performance on light search at night was uniformly very good and did not improve with greater intensity. The lights showed little effect on the pedestrian-search task during either day or night.

Color affected both the objective light-search task during the day and the rating of subjective conspicuity during both day and night. The different photopic photometric values for different colors that are currently specified by the SAE are approximately consistent with these findings, but there appear to be some discrepancies, particularly at night. More data on color may be useful in reviewing those specifications.

Although the original report provides much more detail on this issue, it basically boiled it down to three basic recommendations based on the results of the experiment and on previous results in existing literatures:

1.   Use different intensity levels for day and night.

2.   Make more use of blue overall, day and night.

3.   Use color-coding to indicate whether or not vehicles are blocking the path of traffic.

The strongest findings in research concern the differences between night and day in performance on the light and pedestrian-responder-search tasks. These effects are consistent with the common experience that emergency warning lights are far more visually impressive in the generally dark context of night than against the much brighter context encountered during the day. However, in order to make the best use of warning lights under all conditions, it is important to quantify these differences and the current research results at least begin that effort. For the range of intensities and the flash pattern used in the report, nighttime performance in locating the warning lights was not affected by intensity. Although the older participants made a large number of errors, all participants appeared to be performing as well as possible, at least in the sense that greater stimulus intensities would not have helped. In the daytime, however, the higher intensity level of each of the four colors led to improved performance, indicating that even for the very high range of intensities used in this experiment, visual performance in the search task can still improve. The large overall difference in performance between day and night on the light-search task (853 versus 473 ms) is consistent with that finding, although the very high ambient light levels encountered in the daytime probably make it impossible for any practical warning light to achieve in daytime anything close to the conspicuity levels that most warning lights have at night.

Similarly, reaction times and error rates for the pedestrian-search task at night were substantially worse than during the day. However, the lighting situation was unfavorable to the retroreflective markings, both in terms of the amount of light on the markings and in terms of observation angles and different situations might result in near-daytime levels of performance for pedestrian-responder search. For at least the older group of participants, there appeared to be a measurable negative effect of the flashing warning lights on their ability search for pedestrian responders at night. During the day, performance on the pedestrian-responder-search task appeared to be unaffected by the warning lights, as was expected given the relatively reduced effectiveness of the warning lights in daylight.

There was no difference in performance for the black versus yellow turnout gear either in the day or night. This was expected at night, because under the night lighting conditions, only the retroreflective markings were relevant, and the only difference between the black and yellow turnout gear was in the background material. In daytime, the yellow turnout gear had considerably higher luminance, although, at least for the conditions of this experiment, the difference did not affect visual search for the pedestrian responder.

As was expected, color had effects on both objective search performance and subjective rating of conspicuity. During the daytime, there were marked differences in light-search performance for the different colors beyond the effects that could be attributed to intensity. Researchers interpolated results to determine intensity levels of each of the four colors that corresponded to a single value of reaction time. They found that those levels were at least in rough correspondence to the photometric requirements currently specified in SAE J595. The main exception was that red was less effective in the search task than would be expected based on the SAE requirements. The reaction-time data suggested that blue was very effective in aiding the search task, even in daytime. This is consistent with the SAE requirements, but goes against some statements that have been made about the effectiveness of blue in the daytime. It has often been said that blue is very effective at night (consistent with the idea that the blue-sensitive rod photoreceptors are strong contributors to driver vision at night), but that blue lights provide weak stimuli in daytime.

Subjective ratings of conspicuity were also affected by color, beyond the differences that could be accounted for by differences in intensity. Researchers modeled the effects of color on subjective ratings by determining the levels of intensity for each color that corresponded to a single response level (in this case, a certain value for conspicuity rating). The daytime results are consistent with the SAE J595 requirements, but are inconsistent with the results from the search task. The main discrepancy is that red is subjectively rated as more effective, relative to the other three colors, than it appears to be in the search data. However, there is a reasonably high overall similarity between the effects of color on subjective ratings of conspicuity and the objective effects on reaction time in the light-search task in daytime. The nighttime subjective ratings show a strong difference between red and blue, with red being rated less conspicuous than white, and far less conspicuous than blue. These results are qualitatively consistent with a shift from photopic toward scotopic vision between the daytime and nighttime conditions. They are inconsistent with the current SAE recommendations that are meant to apply to both nighttime and daytime conditions. However, the new results are from a limited range of conditions and it was not possible to quantify the effect of color on the objective search task at night.

To view and download the entire "Effects of Warning Light Color and Intensity on Driver Vision" report, go to the SAE website at: www.sae.org/standardsdev/tsb/cooperative/warninglamp0810.pdf

European Concepts in Roadway Scene Equipment and Practices

The information covered to this point in this chapter has focused on common equipment and practices used for TTC in the United States. This is important because of the mandates set forth in the MUTCD, NFPA standards, and other pertinent regulations. However, emergency response organizations in other parts of the world use different equipment and practices for the same purposes. Though some of these alternatives may not currently meet all the requirements of United States regulations, there is much we can learn from these foreign services. Information on some of these practices is provided here for the sake of furthering research and study of new methods and equipment that can be used to improve responders' safety.

Figure 3.25. It is common for European fire apparatus to have entire compartments dedicated to traffic-control devices.

European fire departments have historically tended to take a more aggressive and participative role in traffic management around roadway emergency scenes than have U.S. fire departments. This may be due to more traffic congestion and typically higher speeds. Fire apparatus in countries such as the United Kingdom and Germany carry considerably more highway safety equipment on their fire apparatus than do typical U.S. fire departments (Figure 3.25).

Figure 3.26. Traffic carts are affixed to the rear of the apparatus.

Figure 3.27. Once the cones have been deployed, the traffic arrow may be placed into service.

Figure 3.28. These traffic paddles are used in the same manner as traffic cones would be deployed.

One innovation used by many German fire departments is a hand cart that carries a variety of traffic-control equipment (Figure 3.26). The cart allows one firefighter to easily deploy a significant amount of traffic safety equipment. A container in the bottom of the cart carries traffic cones, flat traffic paddles that are commonly used throughout Europe, and flashlights or traffic wands. The cart itself is equipped with a large, amber flashing arrow that can be pointed in either direction or simply placed in a flashing alert mode (Figure 3.27). The arrow is powered by a battery on the cart that allows the arrow to be operated for up to 8 hours without recharging.

Figure 3.29. Examples of lighted traffic cones.

The flat traffic paddles that are often carried on these carts are also commonly carried directly on the apparatus. These paddles have weighted bottoms and are used in the same manner as traffic cones or tubular markers. They have a combination of orange stripes and retroreflective white stripes. Many of them are also equipped with LED flashers (Figure 3.28). The advantage of these paddles over cones and tubular markers is the fact that they fold flat and take up considerably less storage space.

Another device commonly found on European apparatus is battery-operated traffic cones that are back-lit for night operations (Figure 3.29). In addition to having retroreflective stripes on the cones, they glow brightly during night operations, making them much more visible before coming into range of the approaching vehicle's headlights.

European fire and police vehicles commonly tow large arrow-board trailers behind their vehicles when responding to roadway incident scenes (Figure 3.30). These may be placed for advanced warning or as directional signals at the beginning of a taper or lane change. It is also common for standard fire apparatus, such as pumpers and rescue units, to be equipped with large, lighted arrow boards (Figure 3.31).

Figure 3.30. This trailer contains a large arrow board.

Figure 3.31. Large lighted traffic arrow boards are commonly affixed to apparatus in Europe.

Many of these ideas have application for U.S. emergency responders and should be considered for implementation into equipment inventories and SOPs.

Recommendations for Roadway Safety Equipment

- Consider the use of ACN systems on emergency vehicles, especially in rural areas.

- Ensure all channelizing devices meet applicable requirements.

- Ensure flaggers, if used, meet MUTCD qualifications.

- Require members to wear ANSI II, ANSI III, or ANSI 207 public safety vest-compliant personal protective equipment (PPE) when conducting highway operations.

- Mark apparatus with conspicuous, contrasting colors.

- Use contrasting-colored restraints and enforce mandatory use during any response.

- All emergency and roadway response vehicles should carry five traffic cones. This is required by NFPA 1901 for new fire apparatus.

- Ensure all vehicles are equipped with lighting systems that can be dimmed or turned off to avoid blinding other motorists.

- Consider equipping each vehicle with one advanced warning sign.

Chapter 4 Setting Up Safe Traffic Incident Management Areas

The previous chapter of this document gave some basic background on concepts of effective temporary traffic control (TTC) and the equipment used to achieve this. This chapter will focus on using this information to set up a safe and effective traffic incident management area (TIMA).

The "Manual on Uniform Traffic Control Devices for Streets and Highways" (MUTCD) defines a traffic incident as "an emergency road user occurrence, a natural disaster, or other unplanned event that affects or impedes the normal flow of traffic" (Section 6I.01). Traffic incidents are divided into three general classes of duration:

1. Minor—expected duration under 30 minutes.

2. Intermediate—expected duration of 30 minutes to 2 hours.

3. Major—expected duration of more than 2 hours.

A TIMA is an area of a highway where TTCs are installed, as authorized by a public authority or the official having jurisdiction of the roadway, in response to a road-user incident, natural disaster, hazardous material spill, or other unplanned incident (Figure 4.1). It is a type of TTC zone and extends from the first-warning device (such as a sign, light, or cone) to the last TTC device, or to a point where vehicles return to the original lane alignment and are clear of the incident. All responders should be trained to work next to motor vehicle traffic in a way that minimizes their vulnerability to being struck by passing traffic. Those having specific TTC responsibilities should be trained in TTC techniques, device usage, and placement.

The primary functions of TTC at a TIMA are to inform road users of the incident and to provide guidance information on the path to follow through the incident area. Alerting road users and establishing a well-defined path to guide road users through the incident area will serve to protect the incident responders and those involved in working at the incident scene. It will also aid in moving road users to expeditiously pass or go around the traffic incident and will reduce the likelihood of secondary traffic crashes and preclude unnecessary use of the surrounding local road system. Examples include a stalled vehicle blocking a lane, a traffic crash blocking the roadway, a hazardous material spill along a highway, and natural disasters such as floods and severe storm damage.

Emergency responders do not have to meet all MUTCD requirements for TTC during the initial phase of a highway incident. The MUTCD requirements for TTC beyond the basic cones, flares, or fluorescent pink signs begin 30 minutes after scene arrival. By this time, law enforcement and highway agencies should be on the scene to establish compliant TTC that fully meets at least the MUTCD minimum standards for the extended incident. Fire departments should accept the responsibility for providing a minimum level of traffic-control devices to be carried on each responding apparatus and to direct traffic until law enforcement arrives.

Figure 4.1. A vehicle collision scene is considered a TIMA. Courtesy of Ron Moore, McKinney, TX, Fire Department.

Drivers have a variety of driving-skill levels. Some drive without a license. Some drive extremely slow. Some drive well beyond the speed limit. Some drive visually impaired. Some drive alcohol/drug impaired. And, all of them "rubberneck" the scene instead of focusing on the road. Incident work zones should be set up to provide the best possible protection of the work area and personnel from vehicle traffic and any other potential hazards.

Establishing the Work Area

As with any type of incident, the key to a successful roadway incident scene operation is getting the incident off to a good start and then building upon that foundation. Getting the incident off to a good start actually begins at the time of dispatch. It is important that the correct units are dispatched to the initial call for help. The units that are dispatched to the incident must then take the safest and most expedient route of travel. Along the way, personnel should draw upon their previous experiences and knowledge of the reported incident location to determine possible appropriate courses of action once they arrive.

This section examines some of the basic steps in getting the roadway incident off to a good start. This includes proper positioning of the initial-arriving vehicles, performing an effective sizeup of the incident, and determining the traffic control procedures that will be required.

Emergency Vehicle Placement

Effective and safe management of a roadway incident scene begins with the arrival and positioning of the first emergency vehicle. From the very outset of the incident, it should be the goal of all responders to protect the incident work area and those who will be operating within this area. According to "Improving Apparatus Response and Roadway Operations Safety in the Career Fire Service," developed by the International Association of Fire Fighters (IAFF) in conjunction with the U.S. Fire Administration (USFA), the emergency vehicle operator has three primary concerns when determining where to park the emergency vehicle on a roadway emergency scene:

Figure 4.2. Emergency vehicles should be used to shield emergency responders from oncoming traffic when they are operating in the incident work area. *Courtesy of Ron Moore, McKinney, TX, Fire Department.*

1. Park the emergency vehicle in a manner that reduces the chance of the vehicle being struck by oncoming traffic.

2. Park the emergency vehicle in a manner that shields responders and the operational work area from being exposed to oncoming traffic (Figure 4.2).

3. Park the emergency vehicle in a location that allows for effective deployment of equipment and resources to handle the incident.

The procedures for performing each of these options will differ depending on the type of incident, the type of road, and the surroundings of the emergency scene. Drivers must be versed in the appropriate positioning procedures for all of the possible environments within which they may be expected to operate.

Operations on Surface Streets

Surface streets range from rural, unpaved roads to busy, urban and suburban avenues. Most often, the tactical needs of the incident will dictate the positioning of the emergency vehicle. However, there are some safety principles that must be followed as much as possible:

*   Park the emergency vehicle off the street in a parking lot or driveway, when possible. This reduces the risk of being struck by a moving vehicle whose driver is not paying attention to the emergency scene.

*   Close the street that the emergency is located on to through traffic. This eliminates the potential of a civilian vehicle driving into the emergency vehicle or responders.

*   Do not block access to the scene for later-arriving emergency vehicles. Oftentimes, crashes occur when one vehicle is parked in a poor position and another attempts to squeeze around it.

Figure 4.3. Always protect the patient-loading area with appropriate equipment and markers.

Figure 4.4. When the incident occurs in an intersection, it may be necessary to shield the work zone from multiple directions. *Courtesy of Ron Moore, McKinney, TX, Fire Department.*

- If the emergency scene is in the street, such as with a vehicle fire or motor vehicle crash, and the street may not be closed to all traffic, park the emergency vehicle in a manner that uses it as a shield between the scene and oncoming traffic. It would be better for a stray vehicle to drive into the emergency vehicle than it would be for it to strike a group of responders.

- On emergency medical services (EMS) calls, use another emergency vehicle to shield the patient-loading area behind the ambulance (Figure 4.3). This area is particularly vulnerable to oncoming traffic. If at all possible, the ambulance should be pulled into a driveway or otherwise out of the route of traffic to reduce the exposure of the loading area.

- Never park the emergency vehicle on railroad tracks. Keep the emergency vehicle far enough away from the tracks so that a passing train will not strike it. Park the emergency vehicle on the same side of the tracks as the incident. This negates the need to stretch hoselines across the tracks or for personnel to be traversing back and forth between each side.

- Position pumping apparatus so that the pump panel is located on the opposite side of the vehicle from oncoming traffic. This will protect the pump operator from being struck by a stray vehicle.

When the incident occurs in an intersection, it may be necessary to shield the incident scene from two or more directions (Figure 4.4). Whenever possible, law enforcement personnel must be used to assist with scene protection and redirection of traffic at these incidents. If sufficient law enforcement personnel are not available to adequately redirect traffic and protect the scene, additional fire companies may be dispatched and their apparatus used to shield the scene. The additional personnel that respond with the extra apparatus can be used to assist with onscene tactical operations or to perform flagging and other scene protection duties.

Operations on Highways

Some of the most dangerous scenarios faced by emergency responders are operations on highways, interstates, turnpikes, and other busy roadways. There are numerous challenges relative to apparatus/vehicle placement, operational effectiveness, and responder safety when dealing with incidents on limited-access highways.

Simply accessing the emergency scene on a limited-access highway can be a challenge to emergency responders. Fire apparatus and other emergency-response vehicles may have to respond over long distances between exits to reach an incident. In some cases, they will be required to travel a long distance before there is a turnaround that allows them to get to the opposite side of the median. Emergency vehicles must not be driven against the normal flow of traffic unless it can be confirmed that police units or highway department officials have closed the road.

Figure 4.5. Place the apparatus between oncoming traffic and the work zone. *Courtesy of Ron Moore, McKinney, TX, Fire Department.*

The driver/operator must use common sense when responding to an incident on a highway or turnpike. A fire apparatus usually travels slower than the normal flow of traffic, and the use of warning lights and sirens may create traffic conditions that actually slow the fire unit's response. Some fire departments have standard operating procedures (SOPs) that require the driver/operator to turn off all warning lights and audible warning devices when responding on limited-access highways. The warning lights are turned back on once the apparatus reaches the scene. However, as will be discussed later in this section, only select warning lights must be used to prevent the blinding of oncoming civilian drivers.

It is important that all emergency response personnel have a good working relationship and compatible SOPs when operating at highway incidents. At a minimum, at least one lane next to the incident lane must be closed. Additional or all traffic lanes may have to be closed if the extra lane does not provide a safe barrier. More detailed information on lane closures is covered in the next portion of this section.

Fire apparatus and other emergency vehicles must be placed between the flow of traffic and the personnel working on the incident to act as a shield (Figure 4.5). Fire apparatus must be parked at an angle so that the operator is protected from traffic by the tailboard. Front wheels must be turned away from the responders working highway incidents so that the apparatus will not be driven into them if struck from behind. Also, consider parking additional emergency vehicles 150 to 200 feet behind the shielding vehicles to act as an additional barrier between responders and the flow of traffic (Figure 4.6).

Figure 4.6. Additional blocking vehicles should be stationed 150 to 200 feet apart.

All responders must use extreme caution when exiting their vehicles so that they are not struck by passing traffic. The firefighters must only mount and dismount the apparatus on the side opposite of flowing traffic whenever possible. Responders in other types of vehicles that do not allow exiting from either side must be especially careful when exiting on the exposed side of the vehicle. Similarly, personnel are extremely vulnerable to being struck by motorists if they step back beyond the protection offered by properly spotted apparatus.

Similar precautions must be taken when positioning other emergency vehicles at a roadway incident scene. Law enforcement vehicles should be positioned so that they provide a barrier and visual warning between oncoming traffic and the incident work zone. Ambulances should be positioned so that their patient-loading areas are protected from approaching traffic.

Emergency-Vehicle Warning Lights

The use of emergency-vehicle warning lights needs to be discussed, as it is something of a two-edged sword. While it is clear that some lighting is necessary in order to warn approaching motorists of the presence of emergency responders, it is also suspected that too much or certain types of lighting can actually increase the hazard to personnel operating on the scene, particularly during nighttime operations.

Two critical issues related to night visibility are color recognition and glare recovery. Because many emergency vehicle warning lights are red, it is important to remember that as the human eye adapts to the dark, the first color to leave the spectrum is red. Red tends to blend in to the nighttime surroundings.

Vision recovery from the effects of glare depends on the prevailing light conditions. Vision recovery from dark to light takes 3 seconds; from light to dark takes at least 6 seconds. A vehicle traveling at 50 miles per hour (mph) covers approximately 75 feet per second, or 450 feet in the 6 seconds before the driver fully regains night vision. This is extremely important when operating on roadways at night, especially on two-lane roads. Headlights on the apparatus that shine directly into oncoming traffic can result in drivers literally passing the incident scene blind, with no sense of apparatus placement.

Wearing protective clothing and/or American National Standards Institute (ANSI)-compliant traffic vests does not improve the ability of the blinded driver to see personnel standing in the roadway. Studies show that the opposing driver is completely blinded at two-and-a-half car lengths from a vehicle with its headlights on.

Figure 4.7. Properly-positioned floodlights will not impair the vision of approaching motorists.

Studies have also shown that strong stimuli, such as the combination of lights, light colors, and varying degrees of reflection and flashes, hold central gaze and drivers tend to steer in the direction of gaze. This has been termed the "moth effect" and is one aspect researched in a study on emergency-vehicle lighting conducted by the USFA and Society of Automotive Engineers (SAE). Other studies have also shown that this visual attraction is further accentuated when the driver is under the influence of drugs and alcohol.

To reduce the potential negative impacts as a result of glare, headlights and fog lights should be shut off at night scenes. Floodlights should be raised to a height that allows light to be directed down on the scene (Figure 4.7). This can reduce trip hazard by reducing shadows and reduces the chance of blinding oncoming drivers. Many highway safety specialists believe that the rear lights on emergency vehicles parked at a roadway scene should be amber. Some fire departments have moved toward the use of all amber warning lights when parked on the roadway during nighttime operations (Figures 4.8a and 4.8b). In some cases, the vehicles are equipped with interlocks that automatically shut off all nonamber warning lights when the parking brake is set.

Section 6I.05 of the MUTCD addresses the use of warning lights as follows:

> "The use of emergency-vehicle lighting (such as high-intensity rotating, flashing, oscillating, or strobe lights) is essential, especially in the initial stages of a traffic incident, for the safety of emergency responders and persons involved in the traffic incident, as well as road users approaching the traffic incident. Emergency-vehicle lighting, however, provides warning only and provides no effective traffic control. The use of too many lights at an incident scene can be distracting and can create confusion for approaching road users, especially at night. Road users approaching the traffic incident from the opposite direction on a divided facility are often distracted by emergency-vehicle lighting and slow their vehicles to look at the traffic incident posing a hazard to themselves and others traveling in their direction."

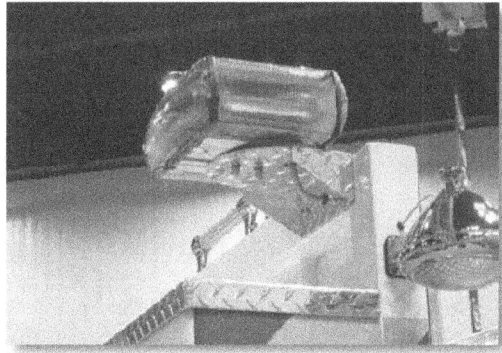

Figure 4.8a. Some agencies use all amber lighting when parked on the roadway.

Figure 4.8b. Some agencies use all amber lighting when parked on the roadway.

Emergency-vehicle lighting can be reduced if good traffic control has been established. If good traffic control is established through placement of advanced warning signs and TTC devices, responders can perform their tasks with minimal emergency-vehicle lighting. This is especially true at major incidents that involve a number of emergency vehicles. Departments should review their policies on emergency-vehicle lighting, especially after a traffic incident scene is secured, with the aim of reducing the use of vehicle lighting as much as possible while not endangering those at the scene. Special consideration should be given to reducing or extinguishing forward-facing vehicle lighting, especially on divided roadways.

Exiting the Apparatus

All responders should have full protective clothing and ANSI-compliant traffic vests as indicated before exiting the apparatus. Check for approaching traffic before exiting. Personnel should exit on the nontraffic side of the vehicle whenever possible. (This may not be possible in apparatus with nonpassthrough jumpseat designs.) Personnel should remember to look down to ensure that any debris on the roadway will not become an obstacle, resulting in a personal injury.

If it is necessary to move around a corner while working at the scene, personnel should move along the downstream, protected side of the apparatus. Stop at the corner of the vehicle and check approaching traffic. Constantly monitor traffic while getting whatever equipment is necessary and moving back to the protected side of the vehicle.

Determining the Magnitude of the Incident

Performing an initial incident sizeup before exiting the vehicle is a primary function of any Incident Management System (IMS). Typically, the lead person on the first-arriving unit will perform an initial sizeup of the incident that is found. This commonly includes an evaluation of the current situation, the actions that will be required to mitigate the situation, and the resources that will be needed to support those actions.

Historically, however, fire department personnel have focused their sizeup solely on handling the incident that is found, be it a collision, injury, or fire on the roadway. What has often gone unconsidered is the impact on traffic and the safety situation this may cause responders on the scene. The MUTCD requires initial responders to determine the magnitude of the incident, the estimated time duration that the roadway will be blocked or affected, and the expected length of the vehicle queue (backup) that will occur as a result of the incident. This information must then be used to set up appropriate TTC measures to handle the incident. Keep in mind that for every 1 minute a lane of traffic is blocked, 4 minutes of backup is developed. This fact emphasizes the need for a quick, accurate sizeup and the implementation of appropriate TTC procedures as soon as possible.

Note that this requirement is not necessarily placed on the actual first-arriving responder, but simply someone in the first group of responders to the incident. In many cases, the fire department will be the first actual emergency agency on the scene, but later-arriving law enforcement or U.S. Department of Transportation (DOT) response units may handle the evaluation of traffic needs. It is just important that **someone** perform this task as soon as safely possible. The goal will be to classify the incident into one of the three categories described in the first part of this chapter: minor, intermediate, or major incidents.

Minor Incident

Minor traffic incidents are typically disabled vehicles and minor crashes or fires that result in lane closures of less than 30 minutes. Diversion of traffic into other lanes is often not needed or is needed only briefly. Traffic control is the responsibility of onscene providers, since it is not usually practical to set up a lane closure with traffic-control devices. If possible, these incidents should be removed from the roadway. An example would be a vehicle collision where there are no injuries and the vehicles are in a drivable condition. All information can be exchanged in a safe location rather than being exposed to oncoming traffic.

Intermediate and Major Incidents

Intermediate incidents affect travel lanes for a period of 30 minutes to 2 hours and usually require diversion of traffic past the incident (Figure 4.9). Full roadway closures might be needed for short periods during the course of the incident. Major incidents typically involve hazardous materials, fatal vehicular collisions involving multiple vehicles, and natural or manmade disasters and extend beyond the 2-hour mark.

Figure 4.9. This vehicle fire in a tunnel is considered an intermediate incident. *Courtesy of Ron Jeffers, Union City, NJ.*

In intermediate and major incidents, traffic is diverted through lane shifts or detoured around the incident and back to the original roadway. Thought must be given to large trucks, especially when having to detour them from a controlled-access roadway onto local or arterial streets. Large trucks may have to follow a separate route from cars because of weight, clearance, or geometric restrictions. Vehicles carrying hazardous materials might need to follow a different route from other vehicles. Gaining the cooperation of the news media in publicizing the existence of (and reasons for) major TIMAs can be of great assistance in keeping drivers and the general public informed and providing alternate traffic routes.

All traffic-control devices needed to set up the TTC should be available for ready deployment at both intermediate and major incidents. The TTC should include proper traffic diversions, tapered lane closures, and upstream warning devices.

Figure 4.10. A typical TIMA.

Expanding the Work Area

According to the MUTCD, minor traffic incidents (those that last less than 30 minutes in duration) may be handled using the equipment at hand. For fire service purposes, this typically means using the apparatus for blocking and perhaps setting out a few markers or signs. The same can be stated for other emergency response agencies who may arrive first at an incident. If the incident will expand beyond this level or duration, a more formal TTC zone will need to be established. While some fire departments may have all the resources to establish a formal TTC zone, in most cases, it will require cooperation between the fire department, law enforcement, and DOT responders to establish this operation. Regardless of who is involved, the MUTCD is specific on the setup of the TTC operation.

As described in Chapter 3 of this document, responders will need to establish a formal TIMA. The TIMA includes the advance warning area that tells motorists of the situation ahead, the transition area where lane changes/closures are made, the activity area where responders are operating, and the incident termination area where normal flow of traffic resumes (Figure 4.10). The distances for the advance warning and transition areas will differ depending on the speed limit in the area of the incident. Higher speed limits require longer advance warning and transition areas. Table 4.1 contains the appropriate lengths based on the speed limit in the area.

Table 4.1. Manual on Uniform Traffic Control Devices for Streets and Highways Traffic Incident Management Area Distances

Miles Per Hour	2nd Warning Sign (B)	1st Warning Sign (A)	Transition Area Taper	Buffer Space	Work Space	Termination Area Taper
30	100	100	70	625	Length of incident	100 ft per lane
40	350	350	125	825	Length of incident	100 ft per lane
50	500	500	375	1,000	Length of incident	100 ft per lane
60	1,500	1,000	450	1,300	Length of incident	100 ft per lane
70	1,500	1,000	525	1,450	Length of incident	100 ft per lane

Only MUTCD-compliant signs and channelizing devices should be used to set up the TIMA. The manner in which they will be deployed will be dependent upon who is deploying them and from where they are doing so. Devices being deployed from emergency vehicles that are already positioned on the scene will most likely be deployed starting at the incident scene/emergency vehicle and working back towards the transition taper and advanced warning areas. Later-arriving units assigned to establish part or all of the TIMA markings may deploy channelizing devices starting from the advanced warning area and working towards the incident.

Regardless of the manner in which these devices are deployed, the following safety principles should be practiced by all of those people who will be deploying them:

- All personnel must be wearing MUTCD-compliant high-visibility protective garments when placing markers or doing anything else on the roadway.

- Personnel should always face the traffic they are operating within and constantly pay attention to approaching vehicles.

- When possible, have a properly marked emergency vehicle provide blocking between the oncoming traffic and the person(s) deploying channelizing devices as they are being placed.

- Once the channelizing devices have been placed, retreat to the protection of the incident workspace, unless flagging duties are required.

Realistically, the exact spacing between channelizing devices will be somewhat dependent upon the number of devices available and the distance to be covered. They should ideally be placed at 15-foot intervals. If flares are initially used during nighttime operations, they should eventually be replaced with cones or tubular markers. Using flares and cones next to each other at night increases the visibility of cones and the direction for traffic flow.

Incidents that fall into the major incident criteria will require traffic-control resources that are well beyond what most police and fire agencies have readily available. It is in these incidents that particularly close working relationships with DOT officials are important. They will be able to provide more substantial resources such as barricades, shadow vehicles, and improved signage.

Flaggers

In some cases, it will be necessary to use emergency personnel to assist the traffic management process by performing manual direction of oncoming vehicles. In parts of the country that use fire police personnel, this will be their primary function when responding to incidents. In other cases, it will most likely be law enforcement personnel but, in some cases, fire personnel may need to assist. The MUTCD refers to personnel performing these duties as flaggers and they must meet the requirements set out in Chapter 6E of the MUTCD. Personnel who have not been trained per these requirements should never be assigned to perform flagger functions.

When performing flagging duties, the flagger should stand on the shoulder adjacent to the lane being controlled or in the closed lane next to the controlled lane (Figure 4.11). At a spot constriction, the flagger may have to take a position on the shoulder opposite the closed section in order to operate effectively. The flagger should have an identified escape route and be located far enough in advance of workers to warn them of approaching danger by out-of-control vehicles. The flagger should stand alone and be visible to motorists. The flagger should always wear MUTCD-compliant high-visibility garments, have appropriate hand-held traffic-control equipment, and be equipped with a whistle or air horn to warn downstream coworkers of impending danger.

Figure 4.11. Flaggers should stand in the lane adjacent to the flow of traffic.

The distances shown in Table 4.2 show the stopping sight distance as a function of speed. These distances may be used for the location of flaggers, but may need to be increased for downgrades and other conditions that affect stopping distance.

Table 4.2. Stopping Sight Distance as a Function of Speed

Speed* (mph)	Distance (ft)
20	115
25	155
30	200
35	250
40	305
45	360
50	425
55	495
60	570
65	645
70	730
75	820

* Posted speed, off-peak 85th percentile speed prior to work starting, or the anticipated operating speed.
Source: MUTCD, Chapter 6E.

Terminating the Temporary Traffic Control Operation

TTC measures must be left in place until the incident has reached its conclusion and all personnel and equipment that were located at the incident area have departed. Once it is safe to dismantle the TTC operation, this should, in general, be performed in one of two ways. If a shadow vehicle is available to protect personnel who are picking up TTC equipment, the devices may be picked up starting with the advance warning signs and working toward the incident area.

If a shadow vehicle is not available, it may be safer to begin picking TTC devices up at the incident area and work back towards the advance warning signs and area. When doing this, personnel should always face traffic within which they are working. If an emergency vehicle is available to provide a barrier between oncoming traffic and the devices they are picking up, it should be used, and personnel should stay behind the vehicle. Once all of the equipment has been picked up and stowed, personnel and apparatus should leave the area immediately.

Recommendations for Setting Up a Safe Work Zone

- Extinguish potentially blinding, forward-facing emergency-vehicle lighting.

- Make sure floodlights are not shining into the eyes of oncoming drivers.

- Always wear MUTCD-compliant protective garments when operating on the roadway.

- Carry all necessary traffic-control devices on responding apparatus.

- Position the initial-arriving engine in a blocking position to oncoming traffic.

- Establish an adequately sized work zone.

- Always face traffic when deploying TTC devices.

- Make sure that personnel who will be expected to perform flagging duties are properly trained and equipped.

- Limit the number of vehicles at the scene to only those necessary to control the incident and provide adequate scene protection.

- When possible, locate the incident staging area off of the highway.

Chapter 5 Preincident Planning and Incident Command for Roadway Incidents

In order to effectively and safely manage highway incidents on a regular basis, a two-pronged approach is required. The first prong is preincident planning. Agencies that develop effective preincident plans that emphasize interagency cooperation when responding to highway incidents are more likely to be successful when these incidents occur.

The *Manual on Uniform Traffic Control Devices for Streets and Highways* (MUTCD) notes that in order to reduce response and handling times for traffic incidents, highway agencies, appropriate public safety agencies (police, fire, emergency medical services (EMS), etc.), traffic management organizations, and private sector responders (towing and recovery and hazardous materials contractors) should be included in preincident plans for occurrences of traffic incidents, particularly along major and heavily traveled roadways. These agencies must also establish ways of sharing incident information.

The second prong is effective use of the Incident Command System (ICS) when responding to incidents on the highway (or anywhere else for that matter). The proper use of ICS provides an organized framework under which all responding agencies may efficiently operate. Traffic incidents and secondary incidents are a major cause of traffic congestion. That congestion can be minimized by diverting traffic before large numbers of vehicles are caught in the incident queue and clearing incidents as quickly as possible. The ICS is the most effective and efficient management process for traffic incident management (TIM) and is particularly applicable to the response, clearance, and recovery stages. Its concepts result in reduced clearance times, mitigating the effects of traffic congestion at the incident site, and reduces safety hazards to responders and motorists.

Preincident Planning for Roadway Incidents

To ensure the safety of responders and the best possible outcome for both victims and motorists involved in a highway incident, those working together at the incident must understand each agency's capabilities and work together. Jurisdictional and agency/institutional issues must be resolved before the agencies come together at an incident. This can be accomplished by effective information-sharing and preincident planning.

Sharing Information

In 2004, the National Cooperative Highway Research Program (NCHRP) published Report 520, "Sharing Information Between Public Safety and Transportation Agencies for Traffic Incident Management." The objective of this study was to assess methods, issues, benefits, and costs associated with sharing information between public safety and transportation agencies in support of TIM.

Nine locations were specifically identified for survey. They were selected because the public safety agencies and transportation agencies in those locations were already exchanging information. Table 5.1 shows the locations surveyed and agencies sharing information. It is of particular interest that of the locations studied, only one (Phoenix, AZ) identified the fire department as a key public safety agency for roadway incidents. The remainder of the locations identified law enforcement as the key public safety agency.

Table 5.1. Locations Surveyed and Key Agencies Involved in Sharing Information

Location	Key Transportation Agencies	Key Public Safety Agencies
Albany, NY	New York State Department of Transportation (DOT); New York State Thruway Authority	New York State Police; Albany Police Department
Austin, TX	Texas DOT	Austin Police Department
Cincinnati, OH	Ohio DOT; Kentucky Transportation Cabinet	Hamilton County Department of Communications; Cincinnati Police; Covington Police Department
Minneapolis, MN	Minnesota DOT	Minnesota State Patrol
Phoenix, AZ	Arizona DOT; Arizona Department of Public Safety	Phoenix Fire Department
Salt Lake City, UT	Utah DOT	Utah Department of Public Safety Highway Patrol and Communications Bureau
San Antonio, TX	Texas DOT	San Antonio Police Department
San Diego, CA	California DOT	California Highway Patrol
Seattle, WA	Washington State DOT	Washington State Patrol

Source: NCHRP, (2004). Report 520, "Sharing Information Between Public Safety and Transportation Agencies for Traffic Incident Management."

Four methods of information-sharing were identified. **Face-to-face** involves direct interpersonal activities, usually at joint operations/shared facilities. **Remote voice** includes such things as telephones and mobile radio. **Electronic text** is text messaging by paging, facsimile, or email, and text access to traffic incident-related data systems, including computer-aided dispatch (CAD). It is worth noting here that most existing CAD systems are proprietary and not designed to exchange information with other CAD systems from different vendors. Therefore, public safety and transportation agencies should consider using compatible information systems to establish effective interagency information exchange whenever practical.

Other media and advanced systems include technological methods not addressed in other categories, such as video and other imaging systems and integrated technologies, including advanced traffic management systems and automatic collision notification (ACN) systems. No single method of sharing information was determined to be the best. Characteristics of the local environment and organizations are key factors affecting the success of a method. Table 5.2 shows the types of information-sharing methods used at each of the locations.

Many factors influence multiagency TIM information-sharing. NCHRP identified the broad factors as institutional, technical, and operational. Leaders and organizations must be willing to work within cooperative partnerships and should have frameworks based on formal agreements or regional plans in place to guide day-to-day activities and working relationships at many organizational levels.

Building an effective information-sharing network or maintaining an existing network requires steps to minimize conflict and establish the basis of effective information coordination. Some suggested steps are as follows:

- Establish a working-level rapport with responders from every agency working on incidents.

- Ensure that working-level relationships are supported by standardized operational procedures.

- Create interagency agreements and system interconnections with key involved agencies.

- Institutionalize senior-level relationships among key agencies through policy agreements, interagency organizations, coordinated budget planning, and other processes to ensure operational partnerships will survive changes in political or management leadership.

Although officials from locations surveyed strongly supported sharing traffic incident information and employing multiagency teams to manage traffic incidents, no location could quantify the benefits. There was no data to measure how other TIM practices affected detection, notification, response, clearance time, responder safety, or other areas of performance. The study recommended that a set of performance measures be formulated and data collected and analyzed to promote information-sharing, demonstrate effectiveness, and justify costs. More indepth information on this study and the nine selected locations is available online at: http://gulliver.trb.org/publications/nchrp/nchrp_rpt_520.pdf

Table 5.2. Summary of Information-Sharing by Location

Location	Face-to-Face	Remote Voice	Electronic Text	Other Media and Advanced Systems
Albany, NY	Two colocation sites	Some sharing of public safety radios; some use of commercial wireless service "talk groups"	Shared CAD system	Roadway data; images; video shared remotely
Austin, TX	Colocation site ready to open	Service patrols equipped with local police radios	CAD data to be shared remotely	Closed-Circuit Television (CCTV) control shared with local police
Cincinnati, OH	Transportation center hosts regional Incident Management Team (IMT) operations	Some sharing of public safety radios; some use of commercial wireless service "talk groups"	Shared CAD under development	CCTV and other traveler information shared with public
Minneapolis, MN	Multiple colocation sites	Shared radio system; some use of commercial wireless service "talk groups"	Shared CAD data	CCTV and other traffic management systems are shared
Phoenix, AZ	N/A	Service patrols equipped with State police and DOT radios	DOT data workstations provided to local public safety agencies	CCTV shared with local fire department
Salt Lake City, UT	Colocation site	Shared radio system	Shared CAD data	CCTV and other traffic management systems are shared
San Antonio, TX	Colocation site	Service patrols equipped with local police radios; shared radio system to be deployed	Shared CAD data	CCTV and other traffic management systems are shared
San Diego, CA	Colocation site	Service patrols equipped with local police radios	Shared CAD data	CAD data are posted on traveler information website
Seattle, WA	N/A	Service patrols equipped with State patrol radios; center-intercom system	Shared CAD data	Control of CCTV is shared with State patrol

All locations use standard telephones and facsimile machines for information-sharing. CAD = Computer-Aided Dispatching. CCTV = Closed-Circuit Television. DOT = Department of Transportation. Source: NCHRP.

Developing the Preincident Plan

The basic process for developing a preincident plan has been widely published and taught throughout the various public safety disciplines and will not be repeated in this document. The procedures for developing a preincident plan for roadway incidents are generally no different than for developing them for any other type of incident, with perhaps one major exception. For example, in many cases, when fire departments are developing a preincident plan, they do so within the framework of the fire department only. Preincident plans for single-family dwelling fires or responses to activated fire alarms do not generally take into account other emergency response or similar agencies. The same may be said for police organizations developing plans for handling barricaded suspect situations.

Ideally, the response to roadway incidents will involve a variety of response entities, each with their own specific role to play in the incident. In order for preincident plans to be effective and easily implemented, all agencies that may be covered under the plan must be included in the development of the plan. Many of the typical types of conflict that tend to bubble up at these incidents can be avoided if all the involved agencies understand each other's roles and operating procedures when they respond to roadway incidents.

As stated above, it is not the purpose of this document to detail the basic procedure for developing a preincident plan. Those procedures are commonly covered in other training programs. However, the following is a list of specific concepts that should be applied to preincident planning for roadway incident operations in order for the preincident planning to be effective and for the plans that are developed to be useful.

- Ensure that all agencies or sectors who may respond to roadway incidents are fully involved in the development of the plan.

- Different agencies or disciplines tend to use different procedures or formats/styles for developing preincident plans. Make sure that all of the involved parties agree on a process and format before beginning the planning.

- Make sure that the final plan that is developed is easily understood and implemented.

- Distribute the final plan to all of the involved agencies.

- Each agency involved in the plan should ensure that all of their personnel are trained on their part of the plan and understand their roles.

- Each agency or discipline should make sure that their personnel are at least minimally briefed on the roles and procedures of the others included in the plan. For example, law enforcement personnel should be trained on fire department procedures for positioning apparatus at roadway scenes. This eliminates conflict on the scenes of actual incidents.

- All agencies involved in the plan should participate in training exercises on a regular basis. This ensures that new personnel learn the plan and experienced personnel are refreshed on the plan.

- All of the agencies should meet to review the plan on at least an annual basis. Problems that have been noted since the last review or new situations that need to be addressed can be discussed and the plan modified accordingly. If the plan is modified, all personnel in the affected agencies should be notified of the changes.

Managing Roadway Incidents

Clearly, proper preincident planning and training are important considerations when preparing to respond to roadway incidents. However, when incidents do occur, it will be necessary to effectively apply the principles of sound incident management in order to bring the incident to a safe and satisfactory conclusion (Figure 5.1). All of the agencies that respond to highway incidents must operate under the umbrella of a common command system in order for the incident to run efficiently.

Figure 5.1. All roadway incidents should be operated under the ICS. Traffic safety vests are not required for firefighters wearing a self-contained breathing apparatus (SCBA). *Courtesy of Mike Mallory, Tulsa, OK, Fire Department.*

Prior to the early 1970s, if response agencies had any formal Incident Management System (IMS) in place at all, they were locally-developed systems. The early 1970s saw the development of several model IMSs that would receive wide use throughout the fire service and some other disciplines. The ICS was developed by Fire RESources of California Organized for Potential Emergencies (FIRESCOPE), which is a consortium of agencies that operated together on major incidents in southern California. ICS was eventually adopted by the National Fire Academy (NFA) and most of the Federal fire and disaster response agencies, as well as numerous fire departments throughout the United States.

At about the same time ICS was developed in California, the Phoenix, AZ, Fire Department developed an IMS called the Fireground Command (FGC) System. This system was also used widely throughout the U.S. fire service as a result of extensive lecturing by members of the Phoenix Fire Department. Though these two systems had similarities, there were enough differences that caused problems when agencies ingrained in one or the other tried to work together.

In the early 1990s, an organization called the National Fire Service Incident Management System Consortium (NFSIMSC; later renamed the National Incident Management System Consortium, or NIMSC) was formed for the purpose of developing an IMS that merged elements of ICS and FGC into a single system that all types of response agencies could use. This consortium consisted of representatives of most of the major fire service organizations and Federal agencies. By 1993, the group was in consensus on a merged system and the consortium began to publish a series of model procedure guides that were designed to teach people how to apply this system to specific types of incidents, such as structural fires, urban search-and-rescue incidents, hazardous materials incidents, and roadway emergency incidents. Most agencies that used one of the original systems or the other made necessary adjustments to use the merged system. However, there was still a very large number of response agencies that had failed to adopt either the old or new systems, and they continued to manage their incidents using little or no form of an organized IMS.

Following the tragedy that occurred on September 11, 2001, it became clear to the Federal government that it would be necessary to mandate the use of a single IMS by all response disciplines in the United States in order to effectively manage large-scale emergencies, natural or manmade, that might occur in the future. In Homeland Security Presidential Directive (HSPD-5), "Management of Domestic Incidents," the President of the United States directed the Secretary of Homeland Security to develop and administer a National Incident Management System (NIMS). On March 1, 2004, the Department of Homeland Security (DHS) issued the NIMS to provide a comprehensive national approach to incident management, applicable to all jurisdictional levels across functional disciplines. The NIMS provides a consistent nationwide approach for Federal, State, tribal, and local governments to work effectively and efficiently together to prepare for, prevent, respond to, and recover from domestic incidents, regardless of cause, size, or complexity.

The NIMS establishes standard incident management processes, protocols, and procedures so that all responders can work together more effectively. NIMS components include

- command and management;

- preparedness;

- resource management;

- communications and information management;

- supporting technologies; and

- ongoing management and maintenance.

The NIMS Integration Center (NIC) was established to oversee all aspects of NIMS. This includes the development of NIMS-related standards and guidelines and support to guidance for incident management and responder organizations as they implement the system. The NIC will validate compliance with the NIMS and National Response Framework (NRF) responsibilities, standards, and requirements.

One component of NIMS is a designated ICS to be used on all incidents. With very few minor exceptions, the ICS mandated within NIMS was virtually identical to the merged system that was previously developed by the NIMSC. Because of this, the NIMSC continues to meet and develop model procedures guides that apply NIMS-ICS to particular types of incidents.

In 2004, the NIMSC, in cooperation with the U.S. Department of Transportation (DOT), developed a manual titled *IMS Model Procedures Guide for Highway Incidents*. The purpose of this manual is to introduce the use of ICS and the principles of Unified Command (UC) to all of the agencies involved with roadway incidents. Representatives of most of the roadway-response disciplines participated in the development of this model procedures guide. On the topic of incident leadership, this document states:

> "Rights of assumption of leadership roles can be unclear with highway incidents, especially when they involve several agencies within the same profession (e.g., federal, state, county, and local law enforcement), or several agencies with overlapping jurisdiction (e.g., law enforcement and transportation), or mixtures of both."

Several factors impact leadership issues, including traditions, organizational capabilities, laws or statutes, etc. Leadership issues must be settled at the local level and must be settled in advance to avoid conflict at the incident scene.

The remainder of this chapter is intended to familiarize roadway emergency-response personnel with the principles of applying ICS to roadway incidents. It emphasizes the need for a coordinated response and operations by all of the agencies that respond to roadway incidents. It is likely that a UC structure will be appropriate at many roadway incidents (Figure 5.2). Most of the information in this section is taken, with permission, from the "IMS Model Procedures Guide for Highway Incidents." That document should be consulted directly for more detailed information, case studies, and examples of model systems.

Figure 5.2. Unified Command (UC) is extremely important when multiple agencies or disciplines are operating at an emergency scene. *Courtesy of Ron Jeffers, Union City, NJ.*

Initiating Incident Management

In order for incident management to be successful, effective Incident Command must be established beginning with the arrival of the first emergency responder, regardless of their rank or agency. The first-arriving responder should establish

Incident Command, perform some basic command functions, and take charge of the incident. From the onset of the incident, principles of sound risk management should be integrated into the functions of Incident Command.

Rules of Engagement

Historically, the fire service has been very quick to apply rules of engagement to structure fire, wildland fire, and hazardous materials scenarios. Similarly, law enforcement agencies have rules of engagement for incidents that include hostage situations, tactical situations, and suspicious packages. Transportation agencies routinely set up work zones on busy sections of roadway. However, many of these agencies have not been so quick to apply those same principles to other known hazardous operations, such as emergency response and roadway scene operations. As stated previously in this document, the roadway is one of the most hazardous locations at which emergency responders operate. Therefore, we must apply principles of risk management to these scenes and operations.

The National Fire Protection Association (NFPA) 1500, *Standard on Fire Department Occupational Safety and Health Program*, objective 8.2.2, states that the concept of risk management for the fire service shall be used on the basis of the following principles:

- Activities that present a significant risk to safety of members shall be limited to situations where there is a potential to save endangered lives.

- Activities that are routinely employed to protect property shall be recognized as inherent risks to the safety of members. Actions shall be taken to reduce or avoid hazards and unnecessary risks.

- No risk to safety of members shall be acceptable when there is no possibility to save lives or property.

Other roadway-response agencies may have similar policies depending on the situation being addressed. Rules of engagement apply to all professions and all hazards encountered in conjunction with highway incidents. Therefore, all agencies should adopt common rules for highway incident management. This will greatly assist Incident Commanders (ICs) when considering courses of action. Figure 5.3 shows a template for Model Rules of Engagement as they are applied to roadway emergency scenes. Agencies should consider adopting them into their standard operating procedures (SOPs) and applying them on all roadway incidents.

Highway Incident Model Rules of Engagement
We will balance risks with the benefits of taking any action.

I. We MAY risk our lives a lot, in a caluated manner, for savable lives, or for preventable further injury or death.
II. We WILL NOT risk lives at all, for property or lives that are already lost.
III. We MAY risk lives only a little, in a calculated manner, for salvageable property, or preventable further damage or destruction.
IV. We WILL endeavor to consider the needs of the others in the vicinity.

Engagement Needs Assessment
We will assess the benefits of our planned actions.

I. We WILL consider the likelihood of success of our actions.
II. We WILL consider the benefit we could provide if we succeed.

Engagement Risk Assessment
We will assess the risks of our planned actions.

I. We WILL assess the threats of injury and death to responders and those in their care.
II. We WILL consider the likelihood of threats occuring and their severity.
III. We WILL endeavor to consider threats of property damage or destruction.

Hazards
- Fire and explosion hazards
- Environmental hazards
- Criminal and terrorist threats
- Traffic hazards

Incident Factors
- Condition of crash vehicles
- Scene access and egress
- Environmental conditions
- Evidence
- Risk to vehicle occupants
- Known or probable occupants
- Occupant survival assessment

Responder Capabilities
- Available resources
- Operational capabilities
- Operational limitations
- Training
- Experience
- Rest and rehabilitation

Figure 5.3.

Risk Analysis

Risk assessment is an ongoing process that lasts for the entire incident. The IC should continually re-evaluate conditions and change strategy or tactics as necessary. At a minimum, the risk analysis for a highway incident should consider

Hazards

- Fire and explosion hazards
- Environmental hazards
- Roadway damage

- Criminal and terrorist threats
- Traffic hazards

Incident Factors

- Condition of involved vehicles
- Scene access and egress
- Environmental conditions
- Evidence

- Risk to vehicle occupants
- Known or probable occupants
- Occupant survival assessment

Responder Capabilities

- Available resources
- Operational capabilities
- Operational limitations

- Training
- Experience
- Rest and rehabilitation

Commanding The Incident

The ICS, as mandated by NIMS, provides the mechanism for numerous emergency-response disciplines to work together in an integrated and coordinated manner during major incidents. This section will review the highlights of command structure at a highway scene based on the "IMS Model Procedures Guide for Highway Incidents." The reader should refer to the actual document for indepth information.

Establishing Command

As stated above, the first-arriving responder must assume command of the incident and remains in control until command is transferred or the incident is stabilized and terminated. The initial IC has several options based on the incident type, situation, and department policy.

Initially, an incident may not have obvious, visible indicators of its significance/severity and will require **investigation**. If it is a single responder, they should examine the scene for conditions and necessary actions. If multiple responders arrive at the same time, one of them should assume the role of initial IC and go with the crew to provide assistance and supervision. This is often referred to as the **investigation mode**.

Some situations require immediate **intervention**, calling for direct involvement of the initial responders in initial stabilization actions (Figure 5.4). This may be called the **intervention mode**. The IC's direct involvement should not last more than a few minutes. At the end of that time, a) the situation is stabilized; b) the IC must withdraw to establish an Incident Command Post (ICP); or c) command is transferred to a later-arriving officer.

Figure 5.4. This is an example of an incident that requires immediate intervention by the initial responders. *Courtesy of Bob Esposito, Pennsburg, PA.*

Large, complex incidents, or those with the potential for rapid expansion, require the first-arriving responder to establish immediate, direct, overall **command**. This is referred to as operating in the command mode. When choosing the **command mode**, the IC will do nothing other than command activities until relieved of the IC duties. If the initial IC is part of a larger crew, the IC has several options on what to do with the rest of the crew while operating in the command mode. These include

- giving the crew a tactical assignment and placing them in action;

- assigning the crew to work under the supervision of another supervisor or crew leader; and

- assigning the crew to perform staff functions to assist the IC.

Transferring Command

Transferring command must follow predetermined procedures. Often, the first transfer of command takes place via radio since there are only a few resources committed to the incident. Subsequent transfers to higher-ranking supervisors or leaders from a different agency must be conducted face-to-face at the ICP.

In some extremely complex incidents or critical situations, an inbound resource or supervisor may be advised of the intent to transfer command to that person upon their arrival at the scene; however, command cannot be passed or transferred to any person not on the scene.

When command is transferred, it is important for the two parties involved in the transfer to engage in an effective relay of information. The IC who is being relieved must fully brief the oncoming IC of the resources that are on the scene and the actions that are under way. Command cannot be effectively transferred until the new IC is fully apprised of the incident status and situation.

Command Aids

There are a number of aids that can be used to assist the IC in ensuring that the command process remains orderly and well documented. The need to implement any or all of these aids will be dependent upon the size and scope of the incident, as well as the capabilities of the IC and other responders.

In medium to large, complex, or escalating incidents, it is essential to document resources committed on the scene are documented as to their current location, their assigned Division/Group, and resources available. **Tactical worksheets** provide a standardized format for that documentation and allow for a more effective transfer of command if necessary. There are many commercially available tactical worksheets that can be used, or the agency can design its own based on local preferences and resources. Some agencies have electronic forms of these worksheets that can be used on mobile computers at the ICP.

Progress reports provide the initial and ongoing information critical for the IC to make effective and safe decisions. Progress reports should be provided by the first resources assigned to Divisions (geographic areas) or Groups (functional assignments). It is important to communicate both progress toward objectives and when progress cannot be achieved.

The **Incident Action Plan (IAP)** identifies the strategy, tactics, and resources to manage and control the incident within a specified time. The tactics are measurable in both performance and time. Short-term, simple operations may not require a written IAP. Large-scale or complex incidents need a written IAP for each operational period. The IAP must be assessed for effectiveness and modified as necessary.

Organizational Structure

The ICS organizational structure develops based on the nature, size, and complexity of the incident. The only difference between ICS on a large incident and ICS on a small incident is the method of organizational growth to meet the needs of the incident. Expanding the ICS organization is the sole decision of the IC and is done when it is determined that the initial or reinforced attack is insufficient. Terms and titles used in the ICS organizational hierarchy are defined in Table 5.3.

Table 5.3. Incident Command System Organizational Hierarchy

Title	Description	Example
IC	Individual responsible for managing all incident operations	
Officer	Member of the Command Staff	Public Information Officer (PIO), Safety Officer, Liaison Officer
Section Chief	Member of the General Staff	Operations Section Chief, Planning Section Chief, Logistics Section Chief, Finance/Administration Section Chief
Director	Individual responsible for command of a Branch	Medical Branch Director, Traffic Management Branch Director
Supervisor	Individual responsible for command of a Division or Group	Extrication Group Supervisor, Traffic Control Group Supervisor, North Division Supervisor
Manager	Individual responsible for a particular activity within the incident organization	Staging Area Manager, Rehab Area Manager
Unit Leader	Individual responsible for a particular activity within the Operations, Planning, Logistics, or Finance/Administration Sections	Traffic Control Unit Leader, Supply Unit Leader
Single Resource	Individual or piece of equipment and its personnel that can be used on an incident	Patrol car, engine company, ambulance, roadway service patrol

In most jurisdictions, an **initial response** to a reported highway incident consists of one to five single resources. The first-arriving resource assumes command until the arrival of a higher-ranking officer, at which point, command is transferred. If the initial response resources are insufficient, the IC will initiate a **reinforced response**, which may include special resources from within the agency or through mutual aid.

The basic configuration of command includes three levels: strategic, tactical, and task. The strategic level involves the overall command of the incident. All planning, determining appropriate strategy, and establishing Incident Objectives that are included in the IAP are accomplished at the strategic level. Supervisors direct operational activities toward specific incident objectives at the tactical level. Activities at the task level are normally completed by individual companies or specific personnel (Figure 5.5).

Even a single-unit response involves all three levels of the command structure. For example, the officer assumes command, determines the strategy and tactics, and supervises the crew doing the task. Many incidents involve a small number of resources, such as an engine, ambulance, and chief. In this situation, the IC handles the strategic and tactical levels. Resources report directly to the IC and operate at the task level.

Figure 5.5. Activities at the task level are normally completed by individual companies or specific personnel. *Courtesy of the Phoenix, AZ, Fire Department.*

Complex situations often exceed the capability of one officer to effectively manage the entire operation. Dividing an incident scene into Divisions/Groups reduces the span of control to more manageable units and allows the IC to communicate with an organizational level rather than multiple individual officers.

Before Multibranch Structure

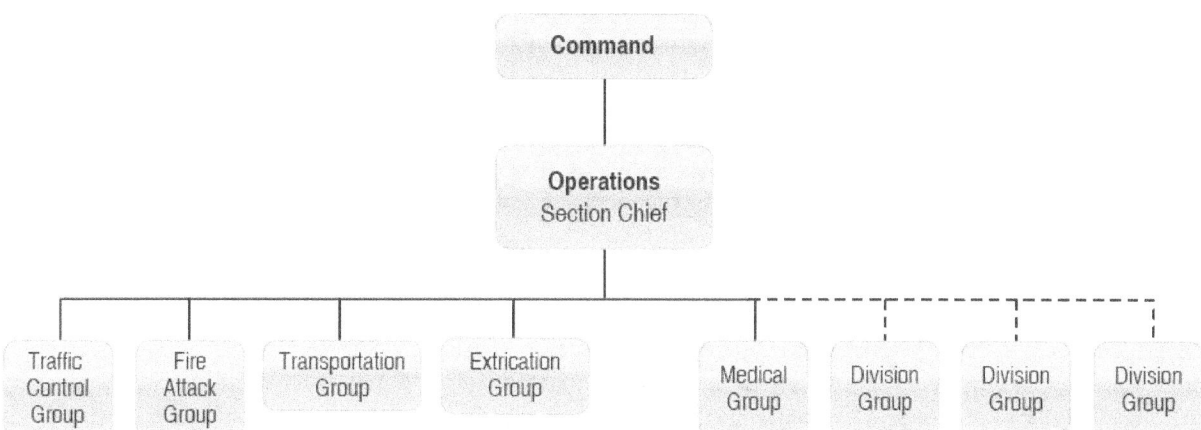

Figure 5.6. This is an example of a command structure that is stretched beyond reasonable means.

Two-Branch Organization

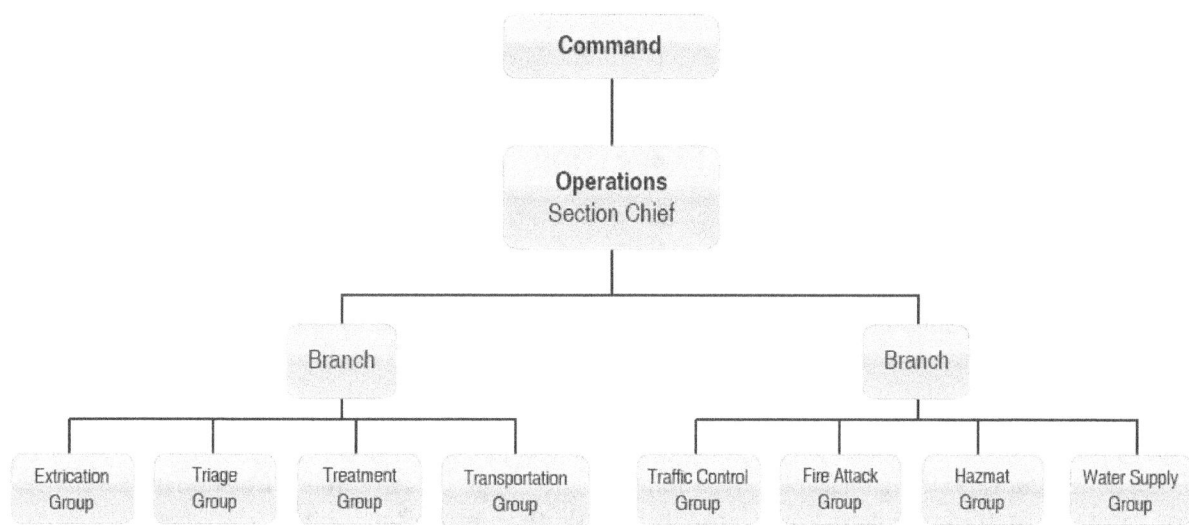

Figure 5.7. A two-branch ICS structure may be warranted at multidiscipline incidents.

Expanding the Organization

When the number of Divisions/Groups exceeds the recommended span of control of three to seven or the incident involves two or more distinctly different operations, the IC may choose to establish a multibranch structure and allocate the Divisions/Groups within those Branches (Figures 5.6 and 5.7).

Some incidents may require a functional Branch structure with each involved department within the jurisdiction having its own functional Branch (Figure 5.8). It is important to remember that resources at multijurisdictional incidents are best managed under the agencies that have normal control over those resources.

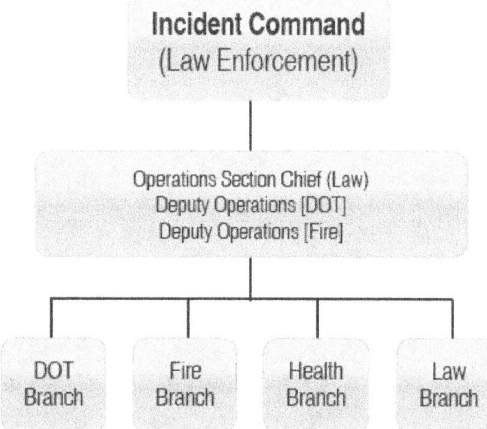

Figure 5.8. An example of an ICS structure that may be warranted at a multidiscipline incident.

Incidents that expand beyond the implementation of a few simple branches in order to manage the assigned resources will typically require the activation of one or more of the four major sections recognized by ICS: Operation, Planning, Logistics, and Finance/Administration. Each of these sections is led by a section chief who reports directly to the IC.

The IC also has the option of appointing three Command Staff positions that report directly to the IC. Command Staff positions are responsible for key activities that are not part of the line organization. Appropriate staff from any involved response organization can fill any of these roles. The PIO is normally the point of contact for the media and other governmental agencies seeking information related to the incident. The Safety Officer assesses hazardous

Figure 5.9. The Operations Section.

and unsafe situations and develops measures for assuring responder safety. The Liaison Officer is the point of contact for representatives from cooperating or assisting agencies and is not directly involved in incident operations. All Command Staff positions can have assistants as indicated by incident complexity. These assistants may represent the various emergency-response disciplines that are involved with the incident.

The **Operations Section** is responsible for the direct management of all incident tactical activities, the tactical priorities, and the safety and welfare of the personnel working in the Operations Section (Figure 5.9). The Operations Section Chief (or simply "Ops Chief") designates an appropriate command channel to communicate strategic and specific objectives to the Branches and/or tactical-level management units. The Operations Section Chief also has responsibility for oversight of Staging Area functions.

The Operations Section is often implemented (staffed) as a span-of-control mechanism. When the number of Branches or Divisions/Groups exceeds the capability of the IC to effectively manage directly, the IC may staff the Operations Section to reduce the span of control, and thus transfer direct management of all tactical activities to the Operations Section Chief. The IC is then able to focus his attention on management of the entire incident rather than concentrating on tactical activities.

Highway incidents often involve the use of aircraft. Aeromedical helicopters may be used to transport patients. Law enforcement may have helicopters in the vicinity and news services may have traffic-reporting helicopters in the area. If the incident is large and prolonged, sightseers in private aircraft may also contribute to air traffic in the area. If aircraft are involved in the operations of the incident, the Operations Section Chief should establish the Air Operations Branch to manage this portion of the incident.

It is important to emphasize that the implementation of an Operations Section is not an automatic event based upon the arrival of the second or third supervisor on the scene. It may be more appropriate to assign later-arriving supervisors to developing Division, Group, or Branch positions first. Experienced supervisors in these positions enhance the command organization and improve the decisionmaking process.

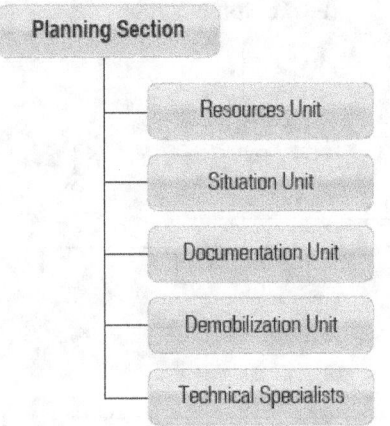

Figure 5.10. The Planning Section.

In some situations, it is more prudent to implement one of the other Section Chiefs before the Operations Section is implemented. For example, a prolonged incident may require the early implementation of a Planning Section before the span-of-control criteria requires an Operations Section Chief.

The **Planning Section** is responsible for gathering, assimilating, analyzing, and processing information needed for effective decisionmaking (Figure 5.10). Information management is a full-time task at large and complex incidents. The automation of traffic management in recent years has greatly increased the amount and quality of information available to traffic managers, enabling them to adjust traffic signals and other controls in reaction to a highway incident. These new traffic management capabilities depend upon receiving information concerning the current situation and also the forecasted duration and extent of incident scene operations. The Planning Section will handle much of this demand for information, working closely in coordination with the Information and Liaison Officers on the Command Staff.

This critical information should be immediately forwarded to Command (or whoever needs it). Information should also be used to make long-range plans. The Planning Section Chief's goal is to plan ahead of current events and identify the need for resources before they are needed. The strategic concerns of the IC need to extend forward with sufficient foresight to cover all of his ICS organization's activities.

Transportation organizations have a great deal of specialized knowledge that can be helpful to the planning function and they should be used as technical specialists by the Planning Section on

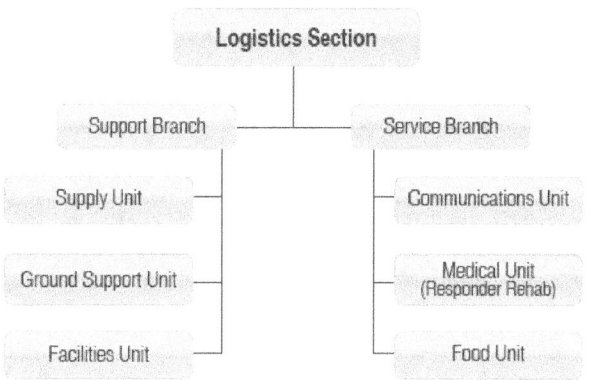

Figure 5.11. The Logistics Section.

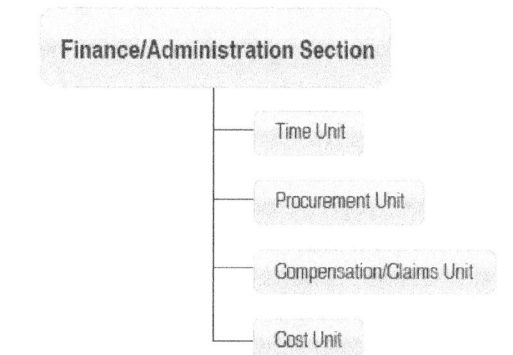

Figure 5.12. The Finance/Administration Section.

complex incidents. These technical specialists are especially helpful when the incident involves more than one mode of transportation, such as rail crossings or transit facilities.

The **Logistics Section** is the support mechanism for the organization. The Logistics Section provides services and support systems, which may be separated into Branches, to all the organizational components involved in the incident, including facilities, transportation, supplies, equipment maintenance, fueling, food services, communications, and responder medical services and rehabilitation. Its organizational breakdown is shown in Figure 5.11.

The **Finance/Administration Section** is established only when involved agencies have a specific need for financial services (Figure 5.12). There are always cost-reimbursement issues with multiagency operations. The designated members of this section are responsible for authorizing expenditures to obtain resources necessary to manage all aspects of the incident.

Unified Command
UC may be appropriate in a) a multijurisdictional incident, such as a collision that crosses city and county lines or b) a multidepartmental incident, such as a collision on an interstate that brings responders from fire, EMS, law enforcement, DOT, and other agencies. The lead agency is determined by the initial priorities. For example, the fire department would be the lead agency if extrication or vehicle fire was involved. As priorities change, the lead agency may change. For example, once all patients have been removed and transported, law enforcement would most likely take over as lead agency. Changes in the lead agency should be accompanied by staffing changes in the Operations Section. Under UC, priorities, strategies, and objectives are determined jointly by the representatives from each agency or jurisdiction.

The importance of an effective UC on major roadway incidents cannot be overemphasized. There are multiple priorities by various agencies on these incidents. Depending on local preferences, it may be desirable to have some form of UC used on most or all roadway incidents. This must be established in SOPs for all response agencies.

Failure to establish UC is often what becomes responsible for conflict between agencies or responders. Some of the concepts associated with using an effective UC are somewhat complex and require preincident planning and training. The concepts surrounding UC exceed what can be covered in this type of document. The *IMS Model Procedures Guide to Highway Incidents* dedicates an entire chapter to this topic. It is highly recommended that agencies consult that document and work those concepts into their SOPs.

Personnel Accountability

The IC is responsible for the overall accountability of personnel operating at the incident scene. Each ICS position is also accountable for all subordinate responders through the chain of command to the IC. However, in large or complex incidents, separate accountability officers may be used.

Currently, there exists no single, nationally recognized or mandated personnel-accountability system in use by any emergency-response agencies. There are several fairly commonly used accountability systems throughout the fire service and other emergency-response agencies. However, they all differ somewhat and are not necessarily interchangeable. Whatever the accountability system used, it must be able to locate every responder at the incident periodically by roll call. When multiple agencies respond, using the combination of each responder's identification number and each agency's name should ensure that responders from all agencies are located (e.g., Green County Engine 1, DOT Response Unit 23).

Emergency Communications

All emergency services should have a standard method for giving emergency message and notification of imminent hazards priority over routine radio communication. NFPA 1221, *Standard for the Installation, Maintenance, and Use of Emergency Services Communications Systems*, is one example of a guideline. It identifies the need to use clear text speech and to have a standard operating guideline (SOG) that uses the term "Emergency Traffic" to clear radio traffic. Clear-text transmissions are also mandated by NIMS-ICS. Any responder, from any response discipline, who is in trouble or subject to an emergency condition can declare Emergency Traffic. At the conclusion of the emergency, an "All Clear" must be transmitted to allow a return to normal radio and incident operations. If response disciplines outside the fire service do not have a standard guide for emergency services communications systems and procedures, NFPA 1221 may be used as a guideline to develop their own procedures.

A signal, such as a truck air horn, can be used in addition to an emergency traffic radio message to signal an ordered evacuation (Figure 5.13). Many agencies use a series of three 10-second blasts of an air horn with a 10-second silence between each series of blasts. If an air horn is used, it is important to make sure the truck is away from the Command Post (CP) to avoid missing radio messages while the horn is sounding.

Transportation Department Roles in the Highway Incident ICS Organization

Transportation departments are important parts in highway incident management and appear frequently on ICS organization charts. The Federal Highway Administration (FHWA) developed the "Simplified Guide to the Incident Command System for Transportation Professionals" to educate transportation personnel and facilitate the integration of this segment into ICS. This guide is available online at: www.ops. fhwa.dot.gov/publications/ics_guide/index.htm#chapt1

Figure 5.13. Vehicle air horns are one manner in which to notify responders of an evacuation.

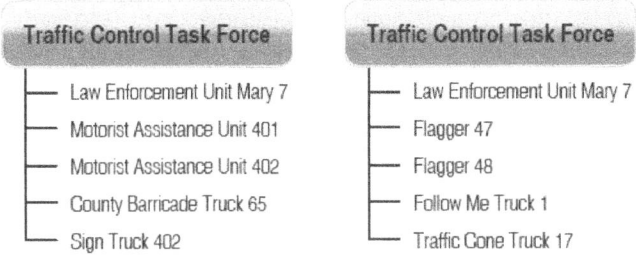

Figure 5.14. Traffic control can be assigned to a task force.

This was also partially the reason that the DOT commissioned the NIMSC to develop the *IMS Model Procedures Guide to Highway Incidents*. The group of subject matter experts that helped to develop that document included representatives of fire, EMS, law enforcement, transportation, and other government agencies.

Traffic control can be easily incorporated into the ICS organization as strike teams, task forces, and traffic management/control groups, divisions, or branches. Strike teams allow the IC to use a significant number of like resources. For example, four police patrol units that are assigned to traffic control could be considered a traffic-control strike team. Task forces organize different types of resources for a specific purpose. An example of this might be two police units, a DOT-response vehicle, and an engine company that are grouped together to set up a traffic incident management area (TIMA) (Figure 5.14).

Traffic-control groups may be formed to consolidate traffic-control functions under a single functional organizational element within the ICS. Traffic management divisions manage a defined geographical part of the highway incident and may be activated to manage traffic movement from separate directions, routes, access points, or intersections. If large numbers of resources are required for this function, there may need to be several groups and/or divisions. This might dictate the need to appoint a Traffic Branch Director to oversee that entire part of the incident organization and operation (Figure 5.15).

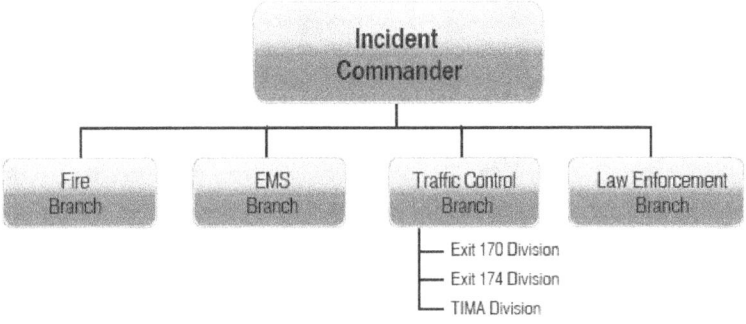

Figure 5.15. When large numbers of traffic-control resources are required at an incident, it may be necessary to establish a Traffic Control Branch.

Organizing the Incident

ICS is applicable to all highway incidents. This section will summarize incidents of increasing complexity based on the *IMS Model Procedures Guide for Highway Incidents*. The reader is strongly encouraged to review the entire publication for a more indepth explanation and examples of highway incident situations with ICS applications.

Prior to Arrival of Response Units

One of the most dangerous times of a highway incident is between when the event occurs and the arrival of the responding units. In addition to the damage, injuries, and/or spills associated with the initial event, traffic is altered with no organized control. Drivers are distracted and often trying to see what has happened rather than watching where they are driving. Further congestion occurs when "good samaritans" stop to help. This situation increases the risk of secondary crashes, resulting in further damage and injuries.

Information regarding the event comes to dispatch from civilians on the scene. It is important that dispatch passes on any additional information that is relevant to responding units to assist in their preparation for managing the incident. This should include items such as the number and types of vehicles involved, number of injured people, and basic information on the possible severity of their injuries, extent of entrapment, fires or hazardous materials involved, and other useful information.

Small Response

Most highway incidents are relatively simple in nature and involve just a few responding units. For example, a vehicle collision with injuries normally warrants a response involving law enforcement, an engine, an ambulance, and perhaps a battalion chief. Depending on the jurisdiction, a DOT-response unit may also be dispatched to assist with traffic control. In this case, the IC assigns companies as they arrive to provide medical care, firefighting and spill control, handle extrications if needed, and manage traffic control and accident investigation. Tasks are prioritized and assigned based on limited resources. This response calls for a simple ICS organization, with all units reporting directly to the IC (Figure 5.16).

Expanded Incident

The ICS organization can be adjusted to deal with additional resources used on an expanded incident (Figure 5.17). Resources are put into common ICS management components to maximize the organizational effectiveness. The basic structure addresses the need for unity of command, a well-defined chain of command, and keeping the span of control manageable. It is generally safe to limit one's span of control to between three and seven subordinates, with an optimum of five.

Reinforced Response

The reinforced response is necessary for the unusual complex highway incidents that require additional resources to deal with further complications such as extreme traffic congestion with the need for more traffic control. Incidents that grow to this level will also typically last for several hours or even more than 1 day. The ICS organization for a reinforced response might look like Figure 5.18.

Additional Considerations

The following are some additional considerations that must be taken into account when operating at roadway emergency scenes.

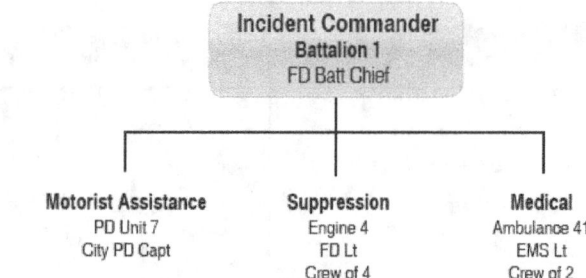

Figure 5.16. The ICS chart for a typical small response.

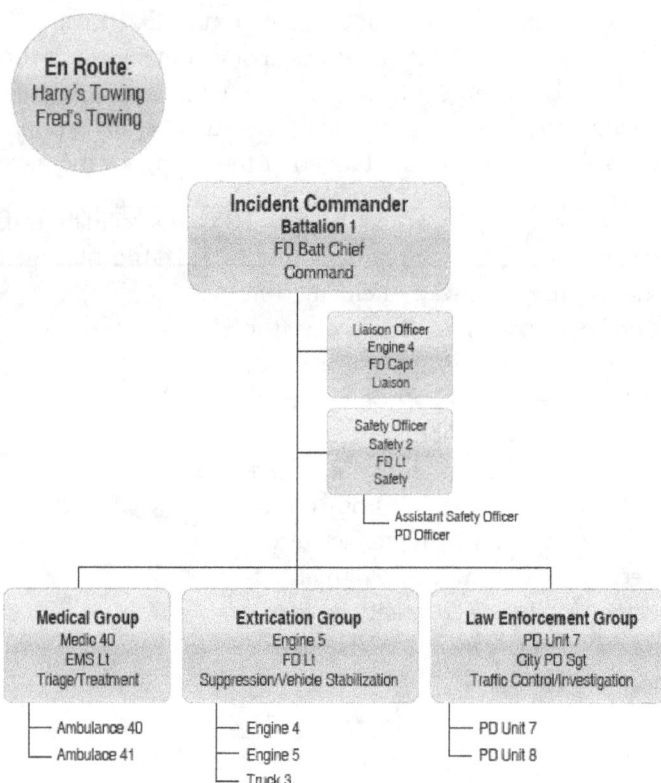

Figure 5.17. This chart shows one example of addressing an expanded incident.

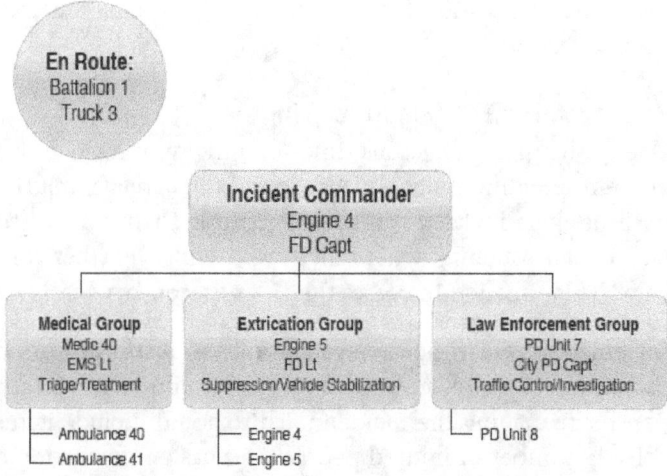

Figure 5.18. Reinforced responses have a more complex command structure when compared to small incidents.

Responder Rehabilitation

The need for responder rehabilitation, usually simply called "rehab," should be considered during the initial planning stages of the emergency response. All supervisors should be aware of the responders in their span of control and ensure their safety and health. In some cases, it is a little more difficult to set up effective rehab at roadway incident scenes than at other, more typical, emergency scenes. Because personnel are often operating while fully exposed to traffic hazards and weather elements, the rehab unit should offer shelter and security (Figure 5.19).

During particularly hot weather, the rehab area should not be set up on the road surface, as this will make it difficult for responders to cool down. In these cases, it would be better to move the rehab area to a grassy area adjacent to the road, under an overpass, or inside a rehab vehicle.

Figure 5.19. Rehabbing emergency responders at long-term incidents is crucial to their well-being. *Courtesy of Ron Jeffers, Union City, NJ.*

Critical Incident Stress Management

All agencies involved in responding to highway incidents should have a method of identifying the incidents that may negatively affect responders and providing appropriate stress-management response. Incidents that involve large numbers of civilian casualties or deaths or those that involve serious injury or death of a responder should result in an automatic critical incident stress management (CISM) response. Since most major highway incidents are multiagency or multijurisdictional, the primary jurisdiction should include the needs of all responders in CISM plans or response.

Recommendations for Managing Highway Incidents

- Develop a formalized TIM information-sharing method between public safety and transportation agencies.

- Manage major traffic incidents using the ICS.

- Consider the use of UC to manage traffic incidents involving multiple jurisdictions or departments.

- Include procedures for operating under UC in preincident plans and practice them on a regular basis.

- Incorporate transportation departments into ICS when appropriate.

Chapter 6 Best Practices and Other Sources of Information
for Effective Highway Incident Operations

An almost endless number of sources exist that fire departments and other agencies which respond to highway emergencies can go to for additional information on how to best handle these incidents. Many of these sources contain detailed information that can only be briefly described here. The purpose of this chapter is to identify some of the more common sources of highway-response information and briefly describe the information available. Responders are strongly encouraged to go to these locations for more information and assistance in forming good plans for dealing with roadway emergency scenes. An extensive list of these websites is located in Appendix B of this document.

The latter portion of this chapter also contains some useful information that can be used by agencies seeking to develop their own standard operating procedures (SOPs) for responding to roadway emergency scenes. The information that will need to be covered in any particular plan will vary depending on location, conditions, and resources. However, the information here will be helpful in showing the kinds of things that should be covered in any plan.

Sources of Information
This section highlights a number of excellent sources for additional information that can be used for agencies seeking to learn more about effective highway incident management operations.

Emergency Responder Safety Institute
Created as a committee of the Cumberland Valley Volunteer Firemen's Association (CVVFA), the Emergency Responder Safety Institute (ERSI) serves as an informal advisory panel of public safety leaders committed to reducing deaths and injuries to America's emergency responders working at roadway incidents. Members of the institute, all highly influential and experts in their fields, are personally dedicated to the safety of men and women who respond to emergencies on or along our nation's streets, roads, and highways. Members of the institute include trainers, writers, managers, government officials, technical experts, and leaders who, through their individual efforts and collective influence in the public safety world, can bring meaningful change.

The ERSI operates a comprehensive informational website at: www.respondersafety.com This website includes breaking news on roadway-related incidents, downloadable training courses and information, roadway incident equipment information, a photo gallery, model SOPs and standard operating guidelines (SOGs), information on "move over" programs, and a large number of links to other related websites. This is an excellent first source of information for people and agencies looking at this topic.

National Traffic Incident Management Coalition
The National Traffic Incident Management Coalition (NTIMC) is a coalition that was launched in 2004 to promote the safe and efficient management of traffic incidents. The NTIMC operates with the support of the American Association of State Highway Transportation Officials (AASHTO). More than two dozen major fire, emergency medical services (EMS), law enforcement, transportation, and other government agencies are members of this coalition.

The purpose of the NTIMC is to work together on strategies for improving congestion relief, responder safety, and domestic emergency preparedness as it relates to roadway emergency scenes. To accomplish this, NTIMC members work together to:

- promote State, regional, and local traffic incident management (TIM) programs;

- promote incident management program standards, best practices, and performance measures; and

- promote incident management program research.

One major accomplishment of the NTIMC was the development of a National Unified Goal (NUG) for TIM in early 2007. The NUG has been ratified by most of the major participants in the NTIMC, as well as other related organizations. The NUG is a unified national policy that encourages State and local transportation and public safety agencies to adopt unified, multidisciplinary policies, procedures, and practices that will dramatically improve the way traffic incidents are managed on U.S. roadways.

The NUG is organized around three major objectives:

- responder safety;

- safe, quick clearance of roadway incidents; and

- prompt, reliable incident communications.

The NUG promotes key strategies related to each theme and accountability to performance targets. Key strategies include development of multijurisdictional, multidisciplinary TIM policies, procedures and training, and development of national, multidisciplinary recommended practices for many operational issues related to TIM. The NUG consists of 18 strategies organized among 4 major topical areas. These are as follows:

Crosscutting Strategies

- **Strategy 1: TIM Partnerships and Programs.** TIM partners at the national, State, regional, and local levels should work together to promote, develop, and sustain effective TIM programs.

- **Strategy 2: Multidisciplinary National Incident Management System (NIMS) and TIM Training.** TIM responders should receive multidisciplinary NIMS and TIM training.

- **Strategy 3: Goals for Performance and Progress.** TIM partners should work together to establish and implement performance goals at the State, regional, and local levels for increasing the effectiveness of TIM, including methods for measuring and monitoring progress.

- **Strategy 4: TIM Technology.** TIM partners at the national, State, regional, and local levels should work together for rapid and coordinated implementation of beneficial new technologies for TIM.

- **Strategy 5: Effective TIM Policies.** TIM partners at the national, State, regional, and local levels should join together to raise awareness regarding proposed policies and legislation that affect achievement of the NUG Objectives of Responder Safety; safe, quick clearance; and prompt, reliable traffic incident communications.

- **Strategy 6: Awareness and Education Partnerships.** Broad partnerships should be developed to promote public awareness and education regarding the public's role in safe, efficient resolution of incidents on the roadways.

Objective 1: Responder Safety

- **Strategy 7: Recommended Practices for Responder Safety.** Recommended practices for responder safety and for traffic control at incident scenes should be developed and widely published, distributed, and adopted (Figure 6.1).

- **Strategy 8: Move Over/Slow Down Laws.** Drivers should be required to move over/slow down when approaching traffic incident response vehicles and traffic incident responders on the roadway.

Figure 6.1. Roadway incident scenes must be managed in an organized manner. *Courtesy of Ron Moore, McKinney, TX, Fire Department.*

- **Strategy 9: Driver Training and Awareness.** Driver training and awareness programs should teach drivers how to react to emergencies on the roadway in order to prevent secondary incidents, including traffic incident responder injuries and deaths.

Objective 2: Safe, Quick Clearance

- **Strategy 10: Multidisciplinary TIM Procedures.** TIM partners at the State, regional, and local levels should develop and adopt multidisciplinary procedures for coordination of TIM operations, based on national recommended practices and procedures.

- **Strategy 11: Response and Clearance Time Goals.** TIM partners at the State, regional, and local levels should commit to achievement of goals for traffic incident response and clearance times (as a component of broader goals for more effective TIM, see Strategy 3).

- **Strategy 12: 24/7 Availability.** TIM responders and resources should be available 24 hours per day/7 days per week.

Objective 3: Prompt, Reliable Incident Communications

- **Strategy 13: Multidisciplinary Communications Practices and Procedures.** Traffic incident responders should develop and implement standardized multidisciplinary traffic incident communications practices and procedures (Figure 6.2).

- **Strategy 14: Prompt, Reliable Responder Notification.** All traffic incident responders should receive prompt, reliable notification of incidents to which they are expected to respond.

- **Strategy 15: Interoperable Voice and Data Networks.** State, regional, and local TIM stakeholders should work together to develop interoperable voice and data networks.

Figure 6.2. Incidents that involve multiple disciplines require a cooperative command structure. *Courtesy of Ron Jeffers, Union City, NJ.*

- **Strategy 16: Broadband Emergency Communications Systems.** National TIM stakeholders (working through the NTIMC) should work together to reduce the barriers to integrated broadband emergency communications systems' development and integration (both wired and wireless).

- **Strategy 17: Prompt, Reliable Traveler Information Systems.** TIM partners should encourage development of more prompt and reliable traveler information systems that will enable drivers to make travel decisions to reduce the impact of emergency incidents on traffic flow.

- **Strategy 18: Partnerships with News Media and Information Providers.** TIM partners should actively partner with news media and information service providers to provide prompt, reliable incident information to the public.

For more detailed information on the NTIMC and/or the NUG, go to their website at: http://timcoalition. org/?siteid=41

U.S. Department of Transportation Federal Highway Administration

The U.S. Department of Transportation (DOT) Federal Highway Administration (FHWA) provides a wealth of information related to safe and effective operations for roadway emergency scenes. The FHWA works are strongly reflective of a Unified Command (UC) and operational effort by all of the various response disciplines who respond to emergencies on the roadway. Following are brief descriptions of some of the sources of information that are available from the FHWA.

Manual on Uniform Traffic Control Devices for Streets and Highways

Much of the previous portions of this document have focused on the contents of the *Manual on Uniform Traffic Control Devices for Streets and Highways* (MUTCD). The DOT/FHWA publishes the MUTCD. This is a large, extensive document, and all emergency-response agencies should refer to it when making TIM plans and preparing to purchase equipment related to this function.

The DOT no longer publishes this document in a printed format, although they do authorize several other publishers to print and distribute it. The DOT does make the entire document available online at http://mutcd.fhwa.dot.gov Once at this site, the user may look at the document online or download portions of the entire document to their own computers.

Best Practices in Traffic Incident Management

"The Best Practices in Traffic Incident Management" report recognizes that TIM is a planned and coordinated program to detect and remove incidents and restore traffic capacity as safely and quickly as possible. This report describes task-specific and crosscutting issues or challenges commonly encountered by TIM responders in the performance of their duties, and novel and/or effective strategies for overcoming these issues and challenges (i.e., best practices). Task-specific challenges may include obtaining accurate information from motorists, accessing the scene, and condemning a spilled load. Crosscutting challenges may include interagency coordination and communication, technology procurement and deployment, and performance measurement. The reported tools and strategies for improving TIM range from sophisticated, high-technology strategies to simple, procedural strategies. This document may be viewed and downloaded from: http://ops.fhwa.dot.gov/publications/fhwahop10050/index.htm

Traffic Incident Management Handbook

The FHWA *Traffic Incident Management Handbook* covers the latest advances in TIM programs and practices across the country and offers insights into the latest innovations in TIM tools and technologies. The new 2010 edition also features a parallel, web-based version that may be conveniently bookmarked, browsed, or keyword-searched for quick reference. Users will find the following topic areas in this handbook:

- **Introduction:** This chapter provides an overview of TIM and sets the context for the 2010 *Traffic Incident Management Handbook* update.

- **TIM Strategic Program Elements:** This chapter details the programmatic structure and institutional coordination necessary for a successful TIM program.

- **TIM Tactical Program Elements:** This chapter describes the full range of onscene operations.

- **TIM Support Program Elements:** This chapter describes the communications and technical aspects of successful TIM programs.

The *Traffic Incident Management Handbook* can be viewed and downloaded at no charge from the following link: http://ops.fhwa.dot.gov/eto_tim_pse/publications/timhandbook/tim_handbook.pdf

Simplified Guide to the Incident Management System for Transportation Officials

This document may downloaded at: http://ops.fhwa.dot.gov/publications/ics_guide/ics_guide.pdf

U.S. Fire Administration Roadway Operations Safety Website

The U.S. Fire Administration (USFA) maintains this page on the USFA website that includes information on all of the work that agency is doing in the area of roadway safety: www.usfa.fema.gov/fireservice/research/safety/roadway.shtm

It also contains links to other related websites and numerous related documents available for viewing and download.

This page of the USFA website provides information on emergency vehicle safety projects and initiatives, many of which also impact on roadway operations safety and response: www.usfa.fema.gov/fireservice/research/safety/vehicle.shtm

National Highway Traffic Safety Administration

The National Highway Traffic Safety Administration (NHTSA) is a division within the DOT focused solely on a broad variety of issues and areas related to safety within the nation's transportation system. The range of information and programs available from NHTSA is exceptionally broad and covers topics that would not immediately seem related to DOT issues, such as the Nation's baseline standards for emergency medical response qualifications. The resources available from NHTSA can be located at: www.nhtsa.dot.gov

Other Examples/Sources of Information

At any given time, there are a number of initiatives related to roadway incident safety underway. Most agencies that are working on these projects are willing to share information with other agencies that are doing the same. The following is a compilation of projects that were in progress or recently completed at the time this document was produced.

Minnesota Traffic Incident Management Recommended Operational Guidelines

www.dot.state.mn.us/tmc/documents/Freeway%20Incident%20Management.pdf

This is the State of Minnesota's protocol for TIM. The purpose of the document is to provide incident responders with uniform guidelines for safe operations at the scene of an incident. The Incident Management Coordination Team has created a document that is easy to read and understand. It lists the roles and responsibilities of each responding agency in clear and simple language and then sets out guidelines for response to typical incidents, including disabled vehicles, crash with property damage only, crash with minor injury, vehicle fire, brush fire (within freeway right-of-way), crash with possible fatality, heavy-duty recovery, and abandoned hazardous materials.

Contact:

Minnesota Department of Transportation Central Office
Transportation Building
395 John Ireland Boulevard
St. Paul, MN 55155
Phone: (651) 296-3000

Strategic Plan for Highway Incident Management in Tennessee

This is a comprehensive look at the issues and needs for transportation incident management from the perspective of the DOT. It is well researched and well written and sets forth the action steps needed to establish inclusive TIM policies and procedures. Particularly interesting is the documentation of the problems that TIM is designed to address. All the stakeholders were included in the planning process. Currently, there is no emergency response manual, although the need for one is identified as an action step to accomplish. The Tennessee DOT does operate a Statewide network of highway service patrol and response vehicles.

Contact:

Office of Incident Management
Tennessee Department of Transportation
Transportation Management Center
6603 Centennial Boulevard
Nashville, TN 37243

Vanderbilt Center for Transportation Research
Box 1831, Station B
Vanderbilt University
Nashville, TN 37235

Emergency Traffic Management in Calgary, Alberta, Canada

Concerns with the safety of responders operating on the roadway and efficient management of highway incidents are not limited to the United States. These issues are of concern anywhere highway systems exist. Several excellent examples of informational documents and guidelines are available from Canadian agencies. The Calgary, Alberta, Fire Department has developed a paper that applies principles of the MUTCD to roadway incidents in their jurisdiction.

This paper examines the logistics of establishing a safe work zone for emergency operations on a highway. It explains how to use traffic cones to create transition zones and lane closures and how to position fire apparatus to protect first responders and those they are working to assist. Graphs illustrate how to create safe zones around bends and on inclines. The document explains terms used in the MUTCD and applies them to emergency operations in a clear and usable manner.

Contact:

Calgary Fire Department
4124 – 11 Street SE.
Calgary, Alberta T2G 3H2 Canada

Nova Scotia Traffic Management Guidelines for Emergency Scenes

www.gov.ns.ca/enla/firesafety/docs/EmergencyRespondersTrafficManagementGuidelines-EmergencyScenes.pdf

This is a manual developed in late 2006 for fire service responders in Nova Scotia. It comprehensively documents how to establish safe work zones for a variety of highway scenarios, including illustrations of how to place cones and position fire apparatus. The manual addresses both career and volunteer firefighters and the issues they must address for safe response. Charts use kilometers per hour instead of miles per hour (mph), so conversions will be needed for United States use. The manual does not address coordination with law enforcement and other highway responders in depth. It does discuss initial response to the scene and proper illumination and signage, as well as appropriate clothing for emergency responders.

Contact:

Public Safety and Office of the Fire Marshal
Nova Scotia Environment and Labor
5151 Terminal Road, 6th floor
P.O. Box 687
Halifax, Nova Scotia B3J 2T8
Toll free: (800) 559-3473 (FIRE)

Standard Operating Procedures

Ensuring the safety of firefighters and other emergency responders while working on the scene of a roadside incident merits the development and use of an SOP. SOPs remind firefighters of actions to follow on the scene and ensure all responders know what actions to expect from others.

This section contains several model procedures that can be used as base material or modified to reflect local conditions and procedures. The first model procedure was originally developed by Battalion Chief/Training Officer Ron Moore of the McKinney, TX, Fire Department for the ERSI. This model procedure has been modified to include other information from a dozen other fire department emergency scene procedures.

The second example is a one-page procedure intended to be a quick reminder of scene-safety survival basics. The last example is the Highway Incident Management Plan from the Hampton Roads area in Virginia. The agencies in this region have long been recognized as leaders in the area of TIM. Portions of their program were also highlighted in the USFA's "Emergency Vehicle Safety Initiative" report.

Additional information on developing effective SOPs may also be found in the USFA document "Developing Effective Standard Operating Procedures for Fire & EMS Departments" located at: www.usfa.fema.gov/downloads/pdf/publications/fa-197-508.pdf

Model Standard Operating Procedure for Safe Operations at Roadway Incidents—Emergency Responder Safety Institute

Purpose

The purpose of this procedure is to provide for the safety of firefighters and other emergency responders on the scene of crashes and other incidents at the roadside and in roadways.

Overview

The first priority for the fire department must be to ensure that its personnel arrive safely at an emergency scene and operate safely at that scene. Operating at roadway incidents is particularly risky due to the hazards posed by moving traffic. Fire personnel must create a safe area to protect themselves and the people they are assisting while taking into account the dangers inherent in working in or near traffic.

In a roadway incident, the fire department's response is only one part of the total mitigation effort. Thus, fire personnel must coordinate their operations with law enforcement agencies, DOTs, and other organizations that may have jurisdiction. The fire department should take the initiative to contact these organizations to work with their personnel in advance of emergencies to determine the roles and responsibilities each will take to make an emergency mitigation effort smooth and effective. Ongoing training involving all organizations will create the cooperation, communication, and trust necessary for safe and efficient public safety service at roadway incidents.

The fire department's primary role at a roadway incident is to safely provide the service needed to stabilize any victims and mitigate the situation without allowing operations to cause additional hazards for passing motorists. For other roadway emergencies such as vehicle fires, the fire must be safely controlled while providing for responder safety. Fire personnel should assume that motorists will be inattentive and/or distracted and gear their operations to account for problems that may arise.

Terminology

The following terms are relevant for roadway incidents and should be used during incidents, in analysis of incidents, and in training for response in or near moving traffic.

Advance warning—Notification procedures used to warn approaching motorists of the need to move from driving normally to driving as required by the temporary emergency traffic-control measures ahead.

Block—Positioning of fire department apparatus at an angle to the lanes of traffic, creating a physical barrier between upstream traffic and the emergency work area. Includes "block to the right" and "block to the left."

Buffer zone—The distance or space between emergency personnel and vehicles in the protected work zone and nearby moving traffic.

Downstream—The direction traffic moves as it travels away from the incident scene.

Flagger—The fire department member assigned to monitor upstream traffic and activate an emergency signal if a motorist does not conform to traffic-control measures and thus presents a hazard to emergency operations.

Shadow—The protected work area of a roadway incident shielded by the block from fire apparatus and other emergency vehicles.

Taper—The action of merging lanes of moving traffic into fewer moving lanes.

Temporary work zone—The physical area of a roadway within which emergency personnel perform their mitigation tasks.

Transition zone—The lanes of a roadway within which upstream motorists must change their speed and position to comply with the traffic-control measures established at an emergency scene.

Upstream—The direction traffic is traveling from as the vehicles approach the incident scene.

Safety Tactics for Fire Personnel
The risk of injury and death when working in and near moving traffic is extremely high. Fire personnel shall use the following tactics to keep themselves safe and reduce their risks:

- Firefighters shall always wear gear with retroreflective trim appropriate to the situation. If turnout gear is not necessary, safety vests with fluorescent retroreflective trim meeting the requirements of ANSI 207, *Standard for High Visibility Public Safety Vests*, with the breakaway option shall be worn.

- At least one firefighter shall always face and be aware of oncoming traffic.

- Firefighters shall use the first-arriving apparatus to establish an initial block to create a temporary work zone (Figure 6.3).

- Firefighters shall exit apparatus on the shadow side, away from moving traffic. If that is not possible, they shall watch carefully and use caution in exiting apparatus. They shall not walk around fire apparatus without taking caution and ensuring that they will be safe in doing so.

- At dawn, dusk, and nighttime, firefighters shall ensure that apparatus headlights, spotlights, and traffic-control strobes that may impair motorists' vision are turned off. Emergency warning lights should be kept to a minimum; more is not better. Amber warning lights are best for all ambient-lighting conditions.

- Working with law enforcement personnel, firefighters shall establish advance warning and adequate transition area traffic-control measures upstream of incidents to allow approaching motorists to reduce travel speeds in the transition

Figure 6.3. First-arriving apparatus should be used to block the incident work zone. *Courtesy of Ron Moore, McKinney, TX, Fire Department.*

zone and pass the incident safely. This includes placing traffic cones and flares at intervals on both the upstream and downstream sides of the incidents.

- A firefighter shall be assigned as flagger to monitor approaching traffic and activate a prearranged emergency signal if a motorist presents danger to firefighters operating in the temporary work zone.

- Firefighters arriving on the scene ahead of responding fire apparatus shall use extreme caution when accessing the emergency scene and while working on the incident scene.

Safety Tactics for Fire Apparatus
- In addition to conveying fire personnel to emergency scenes, fire apparatus shall be used to create safe temporary work zones.

- The first-arriving apparatus shall be angled at about 45° on the roadway with a "block to the left" or "block to the right" to establish a physical barrier between the incident and oncoming traffic.

- If practical, apparatus shall be placed to block the lane of the incident and one additional lane. However, the road should not be closed unless absolutely necessary and with the agreement of law enforcement personnel.

- If practical, apparatus shall be placed so that firefighters can exit on the shadow side and the pump operator can work on the shadow side.

- Apparatus shall be used to block a temporary work zone large enough for all necessary emergency operations.

- Ambulances shall be placed within the temporary work zone downstream of the incident with their loading doors angled away from moving traffic (Figure 6.4).

- If the emergency is at an intersection or near the center of the roadway, two or more sides of the incident shall be protected. The blocking shall be prioritized from the most critical or highest traffic flow side to the least critical. If only one fire apparatus responds, police vehicles shall be used for blocking on the less critical sides.

Figure 6.4. Use the apparatus to shield the ambulance patient-loading zone.

- If apparatus respond to an emergency on a limited-access freeway in the lanes going opposite from where the incident has occurred, they shall use an approved lane to turn around, or go to the next exit and turn around.

- Blocking apparatus shall be positioned in a manner that will prevent it from entering the safe temporary work zone if it is struck by passing vehicles.

Safety Strategy for Incident Command
- The first-arriving Company Officer (CO) and/or the Incident Commander (IC) shall be responsible for ensuring that the emergency operation is conducted in a safe manner.

- The IC shall ensure that fire apparatus provide the necessary blocking to establish a safe temporary work zone. He/She shall establish communications with other agencies on the scene to ensure that the overall response is as smooth and effective as possible. He/She shall ensure that appropriate transition zones are established and marked with cones or flares both upstream and downstream of the temporary work zone.

- The IC shall direct placement of ambulances and parking of additional vehicles to ensure safe medical operations and to ensure that such vehicles do not pose a hazard or a problem to any responding personnel.

- The first-arriving officer and/or IC shall act as scene safety officer until this assignment is delegated.

- The IC shall ensure that the temporary work zone is lighted as needed in such a way that the vision of oncoming motorists is not impaired (Figure 6.5).

- The IC shall manage the termination of the incident as swiftly and effectively as the initial activities. Personnel, apparatus, and equipment shall be removed promptly to reduce exposure to traffic hazards and to minimize congestion.

Figure 6.5. Deploy floodlights in a manner that will not impair the vision of approaching motorists. *Courtesy of Ron Moore, McKinney, TX, Fire Department.*

Equipment

The following equipment shall be available and used appropriately:

- Safety vests meeting the requirements of American National Standards Institute (ANSI) 207, *Standard for High Visibility Public Safety Vests*, with the breakaway option for each emergency responder.

- A minimum of six traffic cones, 28-inch-high fluorescent orange with white reflective striping, as described in the MUTCD.

- Illuminated warning devices such as highway flares or strobes.

- FHWA approved 48x48-inch retroreflective signs stating "EMERGENCY SCENE AHEAD" (with directional arrow overlay) (Figure 6.6).

- Alternating 4-inch fluorescent yellow and red chevron striping shall be installed on the rear vertical surfaces of the apparatus to provide apparatus visibility (Figure 6.7).

Figure 6.6. The MUTCD requires pink reflective signage for deployment in traffic incident management areas (TIMAs). *Courtesy of Ron Moore, McKinney, TX, Fire Department.*

Scene Safety Survival Basics

Roadside Incident One-Page Standard Operating Procedure

- Never trust moving traffic. Behave as if the driver of every vehicle is trying to run you over.

- Wear retroreflective vests or retroreflective clothing while working on the incident scene. Turnout gear is not enough.

- Be situationally aware; remember that you are working on a high-speed roadway just inches or feet away from certain death or injury.

- Use fire apparatus as a shield to protect the incident scene.

- Place ambulances downstream of blocking fire apparatus, if possible, to protect the loading area.

Figure 6.7. Fire apparatus should be equipped with reflective chevron markings on the rear of the vehicle.

- Stage additional ambulances away from the incident scene, if possible, until they are needed.

- Minimize the use of emergency lights at night on the scene. Turn off lights that will blind or confuse oncoming drivers, such as headlights (Figure 6.8).

- Ask law enforcement officers on the incident scene to take an active role in traffic control and scene protection.

- Close the minimum number of traffic lanes while assuring responder safety. Work cooperatively with law enforcement officers on lane closures.

Figure 6.8. The improper use of headlights when emergency vehicles are positioned on nighttime roadway incident scenes can blind approaching drivers. *Courtesy of Ron Moore, McKinney, TX, Fire Department.*

- Clear the scene as soon as possible after patients have been removed and hazards are controlled.

- Post a traffic lookout to alert responders to an out-of-control vehicle.

- Beware of the danger of secondary collisions that will propel vehicles into the incident scene.

Hampton Roads Highway Incident Management Plan
Hampton Roads Highway Incident Management Committee
Hampton Roads, VA

Multijurisdictional Memorandum of Understanding
Highway Incident Management Plan

This Memorandum of Understanding (MOU) is made this 9th day of December, 1999, by and between all Federal, State, county, and city responders to a highway incident in the greater Hampton Roads area (represented by the signatures listed).

The purpose of this plan is to set forth guidance for response to a highway incident in this multijurisdictional area.

It is understood that each responding jurisdictional agency has its own set of operating guidelines and procedures. It is also agreed that each jurisdictional agency recognizes and will implement the UC System should a situation occur that requires such action. This will be accomplished without any agency losing or abdicating authority, responsibility, or accountability.

By way of signature, agency representatives agree to implement the plan through training of their personnel.

Definitions
The Greater Hampton Roads area includes the following counties:

- James City;
- Accomack;
- York; and
- Isle of Wight.

The Greater Hampton Roads area also includes the following cities:

- Chesapeake;
- Franklin;
- Hampton;
- Newport News;
- Norfolk;
- Poquoson;
- Portsmouth;
- Suffolk;
- Virginia Beach; and
- Williamsburg.

Incident—Any situation that impedes the continual flow of traffic. Examples include, but are not limited to, crashes, hazardous materials, fire, medical emergency, etc.

Incident Commander (IC)—Assigned to the first emergency responder arriving at the scene of any highway incident. This role will change as the incident changes.

Responders—Personnel on the scene of any incident.

Traffic-control devices—Items that are used to warn and alert drivers of potential hazards and to guide or direct motorists safely past the hazard(s). May include cones, flares, and signal lights. Advance warning arrow panels (arrow boards) are intended to supplement other traffic-control devices.

Incident safety zone—That portion of the roadway that is closed to traffic and set aside for responders, equipment, and material.

Online video of Hampton Roads Highway Incident Management Plan (HRHIMP) at Center for Transportation Studies, University of Virginia: http://cts.virginia.edu/incident_mgnt_training.htm

Preface

The primary objectives for any operation at the scene of a highway incident are preserving life, preventing injury to any responding personnel, protecting property, and the restoration of traffic flow.

Managing a highway incident and any related problems is a team effort. Incidents range from minor to major with many agencies involved. Each responding agency has an important role to play in the management of an effective incident operation. It is not a question of "Who is in charge?" but "Who is in charge of what?" Each agency present has a part to play. Although the responsibilities may vary from one incident to the next, following are normal practices for agencies in the Greater Hampton Roads area.

Virginia State Police (VSP)—The VSP are the responsible party for responding to traffic incidents on the interstate system. They work in tandem with the respective Federal, county, or city police departments, depending on the circumstances in each situation. The ranking VSP officer is responsible for the incident scene, unless a fire or hazardous material spill is involved; in which case, the ranking fire official is responsible.

Virginia Department of Transportation (VDOT)—Will provide traffic management support, when needed, at an incident scene. VDOT is often relied upon for equipment and personnel for incident support and related activities.

Federal, State, City, and County Law Enforcement Agencies—These agencies may respond to highway incidents in their jurisdiction, depending on the need, availability of personnel, and nature of the incident.

Fire and Rescue Agencies—The determination of the need for fire and rescue services is normally made by the reporting party (call for service). Fire apparatus often respond as a protective measure and additional support.

Towing and Recovery—Will provide the necessary apparatus required for moving and/or removing disabled vehicles from the roadway.

Care of the injured, protection of the public, safety of emergency responders, and clearance of traffic lanes should all be priority concerns of the IC operating at the scene of a highway incident. It is extremely important that all activities that block traffic lanes be concluded as quickly as possible and that the flow of traffic be allowed to resume promptly.

When traffic flow is heavy, small savings in incident-scene clearance time can greatly reduce traffic backups and the probability of secondary incidents. Restoring the roadway to normal or to near normal as soon as possible creates a safer environment for motorists and emergency responders. Additionally, it improves the public's perception of the agencies involved and reduces the time and dollar loss resulting from the incident.

Purpose

The purpose of this plan is to provide incident responders with a uniform guide for safe operations at incidents occurring on the highway system. It is intended to serve as a guideline for decisionmaking and can be modified by the incident responders as necessary to address existing incident conditions.

Emergency operations at the scene of a vehicle accident are the most common occurrences and those with the greatest potential for an unfavorable outcome to personnel. Each year, many significant incidents occur on roadways. Whether it is the interstate highway or a secondary road, the potential for injury or death to any responder is overwhelming.

Response

Emergency responders need to operate safely, making every effort to minimize the risk of injury to themselves and those who use the highway system. Responders operating in the emergency mode need to operate warning devices and follow the guidelines specific to their SOPs.

- Warning lights—Emergency-warning lights should remain operational while responding to and, when necessary, while working at incidents.

- Headlights—Apparatus headlights should be operational during all responses and incidents regardless of the time of the day. Caution should be used to avoid blinding oncoming traffic while on the scene.

- Siren and air horn—When responding as an emergency vehicle, appropriate warning devices will be used in accordance with State law.

Median-strip crossovers marked "Authorized Vehicles Only" shall be used for turning around and crossing to the other travel lanes **only** when emergency vehicles can complete the turn without obstructing the flow of traffic in either travel direction or all traffic movement has stopped. Under no circumstances shall crossovers be used for routine (nonemergency) changes in travel direction.

Use of U-turn access points in "jersey" barriers on limited access highways is extremely hazardous and shall be used only when the situation is necessary for immediate lifesaving measures.

Response on access ramps shall be in the normal direction of travel, unless the IC on the scene can confirm that oncoming traffic has been stopped and no civilian vehicles will be encountered on the ramp.

Shoulder lanes will be used only by emergency vehicles/apparatus. Emergency support vehicles are authorized to use the shoulder lanes only when directed or authorized to do so by the IC.

Arrival

The first emergency responder arriving to the scene of any highway incident will assume the role of IC. The individual assuming that role is subject to change as additional responders arrive at the scene.

If traffic control assistance is required at an incident scene, the IC will request that contact be made to Traffic Management System (TMS) Control (Smart Traffic Center) at (757) 424-9903. By providing a brief description of the situation, VDOT personnel may be dispatched, if not already en route to assist.

Standard practice will be to position response vehicles in such a manner as to ensure a safe work area. This may be difficult to accomplish at incidents on secondary and one-lane roads. Position emergency-response vehicles in such a manner as to provide the safest area possible.

Parking of Response Vehicles

Providing a safe incident scene for emergency responders is a priority at every emergency incident. However, consideration must be given to keeping as many traffic lanes open as possible. Except for those vehicles needed in the operation and those used as a shield for the incident scene, other response vehicles should be parked together (Staging Area). As a matter of routine, the parking of response vehicles should be on one side of the roadway. Parking should be on either the shoulder or median area, if one exists, but not both. Parking response vehicles completely out of available travel lanes greatly assists in the movement of traffic. If not needed to illuminate the scene, drivers should remember to turn vehicle headlights off when parked at incidents.

Recovery personnel are to report to the IC, who will then direct them to a safe, or "Staging" Area.

Onscene Actions

The proper spotting and placement of emergency apparatus is the joint responsibility of the driver and IC. The proper positioning of emergency-response vehicles at the scene of an incident assures other responding resources of easy access and a safe working area and helps to contribute to an effective overall operation. The safety of everyone on the scene is foremost while they are operating, both in emergency and nonemergency situations.

An incident safety zone shall be established, allowing fire and rescue units to position in close proximity of the incident. The responding fire apparatus should be placed back some distance from the incident, making use of it as a safety shield blocking only those travel lanes necessary. In the event that a motorist enters the incident safety zone, the fire apparatus will act as a barrier; and, in the unlikely event that the fire apparatus is moved upon impact, it will travel away from the incident safety zone.

Before exiting any emergency-response vehicle at an incident, personnel should check to ensure that traffic has stopped to avoid the possibility of being struck by a passing vehicle. Personnel should remember to look down to ensure that debris on the roadway will not become an obstacle, resulting in a personal injury. All members shall be in appropriate clothing or traffic vests as the situation indicates.

As soon as possible, the initial-responding unit should position traffic-control devices. Traffic cones assist in channeling traffic away from an incident. Traffic-control devices shall be used whenever responding vehicles are parked on or near any road surface. Placement of traffic-control devices shall begin closest to the incident, working toward oncoming traffic. Taking into consideration the possibility of hazardous materials, traffic-control devices shall be placed diagonally across the roadway and around the incident. This assists in establishing an incident safety zone. When placing traffic-control devices, care should be exercised to avoid being struck by oncoming traffic.

The speed of traffic and travel distance must be considered when establishing an incident safety zone. The following chart provides an example of how traffic-control devices are to be placed.

Posted Speed Limit	Distance
35 mph	100 ft
45 mph	150 ft
55 mph	200 ft
> 55 mph	250 ft and greater

When channeling traffic around an incident, traffic-control devices shall also be used in front of the incident if those devices and the manpower are available.

It is possible to channel traffic around a curve, hill, or ramp, provided the first device is placed such that the oncoming driver is made aware of imminent danger.

Emergency Vehicle Visibility at Night

Glare vision and recovery is the amount of time required to recover from the effects of glare once a light source passes through the eye. It takes at least 6 seconds, going from light to dark, and 3 seconds, from dark to light, for vision to recover.

At 50 mph, the distance traveled during a second is approximately 75 feet. Thus, in 6 seconds, the vehicle has traveled 450 feet before the driver has fully regained night vision. This is extremely important to remember when operating on roadways at night.

The headlights on stopped vehicles can temporarily blind motorists that are approaching an incident scene. Drivers of oncoming vehicles will experience the problem of glare recovery. This essentially means individuals are driving past the emergency scene blindly. The wearing of protective clothing and/or traffic vests will not help this "blinded" motorist see emergency responders standing in the roadway. Studies show that at two-and-a-half car lengths away from a vehicle with its headlights on, the opposing driver is completely blinded.

Low-beam headlights can be used to light an emergency scene using care as to light only the immediate area. Complacency at an incident scene can be hazardous.

Clearing Traffic Lanes

When outside of a vehicle on a major roadway, both civilian and emergency responders are in an extremely dangerous environment. Therefore, it is imperative to take every precaution to protect all responders and those involved at incident scenes. Although positioning emergency-response vehicles to serve as a shield for work areas is a prudent practice, we must remember that reducing and/or shutting down traffic lanes creates other problems and safety concerns. Therefore, it is critical when operational phases are completed that emergency-response vehicles be repositioned to allow traffic to flow on as many open lanes as possible.

Remember that unnecessarily closing or keeping traffic lanes closed greatly increases the risk of a secondary incident occurring in the resulting traffic backup. Five minutes of stopped traffic will cause a 15-minute delay in travel time.

Management of incidents on the interstate system and local roadways requires the expertise and resources of emergency responders, as defined. While the safety of emergency services personnel is of paramount concern for the IC, the flow of traffic must be taken into consideration at all times. The closing of roadways disrupts traffic throughout the area as well as having a significant impact on businesses throughout the region.

Keeping the safety of all personnel in mind, and coordinating the needs with other emergency services, the IC should begin to open any closed lanes as soon as practical.

Chapter 7 Recommendations

The earlier portions of this report emphasize the frequency and consequences associated with first-responder injuries and deaths as a result of incidents that occur on the roadway. Clearly, this is a growing problem that needs to be mitigated. There is no single step that can be taken to significantly improve this problem; it requires a comprehensive approach to better handling of roadway incidents.

Vehicle collisions have both immediate and long-term economic effects on the individual and society. Costs are both direct (those that are the result of the collision and resultant injury/fatality) and indirect (overall cost to society).

The effective use of approved traffic-control devices promotes highway safety and efficiency by providing for orderly movement of all road users. The National Incident Management System (NIMS) Incident Command System (ICS) is the most effective and efficient process for traffic incident management (TIM). Complying with the U.S. Department of Transportation's (DOT's) *Manual of Uniform Traffic Control Devices for Streets and Highways* (MUTCD) and adopting the guidelines contained in the National Incident Management System Consortium's (NIMSC's) "Model Procedures Guide for Highway Incidents" should help enhance emergency responder operational effectiveness, reduce potential liability, and enhance responder safety at roadway emergency scenes.

Based on the research performed to prepare this report, the following additional recommendations are presented to help decrease vehicle-related injuries and fatalities of emergency responders if implemented at the appropriate levels.

1. **Develop a comprehensive database that tracks accidents involving emergency vehicles and any resulting injuries/deaths to both emergency responders and civilians.**

The failure to capture and analyze accurate, useful data on a wide range of issues is an age-old problem in the emergency-response disciplines. Some accurate data is available on firefighter and law enforcement fatalities. However, little reliable data is available on incidents involving injuries or no injuries. Without this data, it is difficult to accurately assess the problems we are facing.

Accordingly, there exists no comprehensive database to determine specific information related to emergency-vehicle collisions. There should be a national repository that collects data from all organizations and allows for retrieval of specific information regarding vehicle collisions responding to/returning from incidents, emergency workers struck by vehicles at the scene, secondary crashes, and civilian injuries/fatalities resulting from collisions with emergency vehicles.

2. **Limit speeds to level that is safe for the vehicle being driven and road conditions on which it is being operated.**

There is a simple old saying that says "speed kills." We certainly know this is true in the emergency services (Figure 7.1). The urgency that we place on responding to emergency calls is often translated into excessive speed during the response. Speeds that are significantly above the posted speed limit are dangerous, especially in fire apparatus and other large emergency-response vehicles. Stopping distances are increased dramatically, and high-vehicle speeds in curves often have negative outcomes. The decision to exceed the posted speed limit should be based on assessing the risk of such speeds with the benefit to those needing assistance. We

Figure 7.1. Operating an emergency vehicle at an excessive speed can have tragic results.

cannot perform effective roadway scene operations if we fail to reach the scene. Furthermore, collisions as a part of an unsafe response add another roadway incident to our load that must be handled. This places even more responders in the roadway.

Each jurisdiction should establish maximum speed policies for the vehicles they operate. One basic way of doing this is establishing a policy requiring that vehicles may not exceed the posted speed limit. If the jurisdiction is within a State that allows emergency vehicles to exceed posted speed limits, the local standard operating procedure (SOP) should not exceed the State's limits. The local jurisdiction may also choose to set speed limits that are below the State requirements if they so desire.

It should be noted that agencies may wish to establish different speed limits for particular types of vehicles. For example, fire departments may wish to establish lower maximum speeds for larger apparatus such as aerial apparatus and water tankers/tenders that are particularly dangerous at higher speeds. The same may be true for law enforcement agencies. Higher-profile vehicles, such as sport utility vehicles (SUVs), may have lower maximum speed limits than standard patrol cars.

3.   Adopt a zero-tolerance alcohol policy and enforce an 8-hour time between alcohol consumption and work.

From 1990 to 2003, there were 17 firefighter fatalities in which alcohol or drugs were a direct factor in the death of a firefighter; the firefighter who died was intoxicated or high, or another firefighter involved in the death was intoxicated or high. Impaired firefighters may be involved in collisions during the response or may take unsafe actions when they arrive on the scene. Between 1997 and 2002, several privately owned vehicle (POV)-firefighter fatalities had blood alcohol concentrations that would be considered legally intoxicated in most States. Departments should adopt the International Association of Fire Chief's (IAFC's) zero-tolerance alcohol policy and enforce the 8-hour time between alcohol consumption and work. Similar data shows that these issues also occur in law enforcement, emergency medical services (EMS), and other emergency-response disciplines. It is also imperative that all emergency-response organizations recognize their members with alcohol abuse problems and provide them with the help that they need.

4.   Equip all emergency vehicles with appropriate traffic control and safety equipment.

All emergency vehicles, including staff and nonemergency-response vehicles, should be equipped with an appropriate supply of traffic control and safety equipment. This includes high-visibility vests, flashlights, and channelizing equipment (Figure 7.2). The amount and type of equipment carried will vary on the responsibilities assigned to the personnel typically riding in that specific vehicle.

Figure 7.2. All emergency vehicles should carry traffic-control equipment.

5.   Ensure all traffic-channelizing devices meet applicable standards.

Channelizing devices used during an emergency incident can include signs, cones, tubular markers, flares, directional arrows, flagger equipment, and related equipment. All of this equipment must meet MUTCD and National Fire Protection Association (NFPA) requirements. All of the equipment should be in good repair and ready for deployment.

6. **Ensure flaggers, if used, are properly trained and meet MUTCD qualifications.**

The MUTCD requires flaggers to have the following abilities:

*   receive and communicate specific instructions;

*   move and maneuver quickly;

*   control signaling devices to provide clear and positive guidance to drivers;

*   understand and apply safe traffic-control practices; and

*   recognize dangerous traffic situations and warn workers in sufficient time to avoid injury.

Teaching these skills should be a basic part of any entry-level training program for members in any emergency-response discipline. It should also be a regular part of recurrent, in-service training for all active personnel.

7. **Require members to wear highly reflective American National Standards Institute (ANSI)/ International Safety Equipment Association (ISEA) 107 Class II, Class III, or ANSI/ISEA 207-compliant public safety vests whenever they operate in the roadway.**

Personnel visibility is critical during highway operations. All apparatus should be equipped with one vest for each riding position on the emergency vehicle and nonemergency vehicles should also carry at least one vest. All members must be required to wear the vests whenever they are operating in the roadway. The only exceptions to the requirement to wear a reflective vest when operating on the roadway are situations in which the personnel are wearing self-contained breathing apparatus (SCBA) or chemical-protective clothing.

8. **Mark the emergency vehicle perimeter with retroreflective striping or markings.**

NFPA 1901, *Standard for Automotive Fire Apparatus*, requires retroreflective striping around the perimeter of new fire apparatus to illuminate the apparatus at night when visibility is limited. Placement of the striping provides an indication of the location and size of the apparatus. NFPA 1901 also requires retroreflective striping inside cab doors to maintain conspicuity and alert them to an open door. NFPA 1901 now requires the use of European-style retroreflective markings on the rear of fire apparatus (Figure 7.3). When feasible, reflective markings meeting the current NFPA requirement should be added to existing apparatus that is still in service. Many other emergency-response disciplines, including law enforcement, EMS agencies, and DOT response units have also increased their use of retroreflective markings on their vehicles in recent years (Figure 7.4). It is highly recommended that all of these agencies implement the use of more conspicuous markings on all of their vehicles.

Figure 7.3. NFPA 1901 requires retroreflective chevrons on the rear of all new fire apparatus. *Courtesy of Jack Sullivan.*

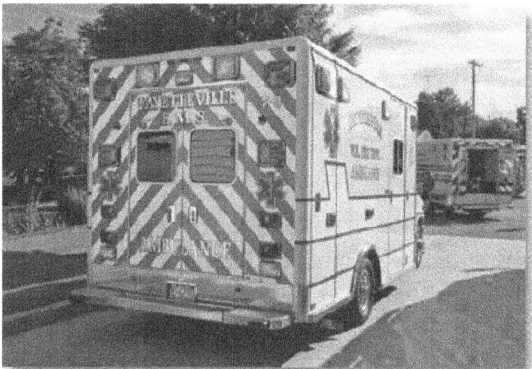

Figure 7.4. Chevrons can be applied to ambulances as well. *Courtesy of Jack Sullivan.*

9. **Extinguish forward-facing emergency-vehicle lighting when parked on the roadway, especially on divided roadways.**

Headlights and fog lights should be shut off at night scenes. Some agencies feel that amber-only lights are safest for the rear of their emergency vehicles. MUTCD states that emergency lighting is often confusing to drivers, especially at night. Drivers approaching the incident from the opposite direction on a divided roadway are often distracted by the lights and slow their response, resulting in a hazard to themselves and others traveling in their direction. It also often results in traffic congestion in the unaffected opposite lane(s) and increases the chance of a secondary collision. If floodlights are being used for nighttime operations, they should be angled downward towards the work area to avoid blinding approaching motorists.

10. **Fire departments should consider the implementation of traffic-safety response units.**

Traffic-safety response units respond to roadway incident scenes and assist other fire personnel on the scene with providing proper blocking and marking procedures. These units are common in the mid-Atlantic region of the United States, but scarcely used in other portions of the country. They are particularly helpful in jurisdictions that have limited law enforcement personnel available onduty. These could be established under the jurisdiction of a fire department safety division or other specified organizational unit.

11. **Position the initial-arriving emergency vehicle in a blocking position to oncoming traffic.**

The blocking position allows the initial responder to survey the scene from inside the emergency vehicle (Figure 7.5). The emergency vehicle should be positioned to ensure a safe work area at least one lane wider than the incident, whenever this is possible. When an incident is near the middle of the street at an intersection, two or more sides may need to be protected. Block all sides of the incident that are exposed to oncoming traffic.

Figure 7.5. Police vehicles may also be used to provide a barrier for the incident work zone.

12. **Establish an adequate size work zone.**

When no fuel, fire, or spill hazards are present, the work zone should extend approximately 50 feet in all directions from the wreckage. If there is a vehicular fire involved, the work zone should extend approximately 100 feet. Low-lying areas should also become extended work zones if the vehicle(s) are leaking fuel, since fumes typically travel downhill and downwind.

13. **Develop a formalized TIM information-sharing method between public safety and transportation agencies.**

Factors involved in developing an effective information-sharing program are institutional, technical, and operational. Implement cooperative partnerships and frameworks based on formal agreements or regional plans to guide day-to-day activities and working relationships. Consider using compatible information systems to establish effective interagency information exchange whenever practical.

14. **Manage major traffic incidents using the NIMS ICS.**

NIMS ICS provides the mechanism for numerous emergency-response disciplines to work together in an integrated and coordinated manner during incidents. It is the most effective and efficient process for TIM and is particularly applicable to the response, clearance, and recovery stages. In addition to improving scene safety, managing a traffic incident using ICS can reduce clearance times, which mitigates the effects of traffic congestion at the incident site.

15. **Consider the use of Unified Command (UC) to manage traffic incidents involving multiple jurisdictions or disciplines.**

UC may be appropriate in a multijurisdictional incident, such as a collision that crosses city and county lines or a multidepartmental incident, as in the case of a collision on an interstate that brings responders from fire, EMS, law enforcement, and DOT. The lead agency should be determined by the initial priorities. As priorities change, the lead agency may change.

16. **Incorporate transportation departments into ICS when appropriate.**

Transportation departments are one of the newer participants in highway incident management (Figure 7.6). Traffic control can be easily incorporated into ICS organization as strike teams, task forces, control groups, or traffic management divisions.

17. **Ensure adequate training on roadway hazards and safety procedures for responders.**

Figure 7.6. Many transportation agencies staff highway safety-response units. *Courtesy of Jack Sullivan.*

Fire departments should increase the amount of training on roadway scene safety provided to personnel who respond to these types of incidents. NFPA 1001, *Standard for Fire Fighter Professional Qualifications*, contains minimum training requirements for entry-level firefighters. At a minimum, this should be followed for all firefighters already on the job. At the time this document was released, the NFPA was also in the early stages of developing a professional qualifications standard for TIM control personnel.

Most other response disciplines also have standardized levels of basic training and/or certification that should also include sufficient training on the topic of roadway scene safety. In areas with greater levels of roadway hazards, additional training should be required. Anyone who will be required to perform flagger duties should be trained as directed by the MUTCD.

Appendix A List of Acronyms and Abbreviations

24/7	24 hours per day/7 days per week
AASHTO	American Association of State Highway Transportation Officials
ALS	Advanced Life Support
ACN	Automatic Collision Notification
ANSI	American National Standards Institute
BLS	Basic Life Support
CAD	Computer-Aided Dispatch
CCTV	Closed-Circuit Television
CDC	Centers for Disease Control and Prevention
CISM	Critical Incident Stress Management
CMS	Changeable Message Sign
CO	Company Officer
CP	Command Post
CVVFA	Cumberland Valley Volunteer Firemen's Association
DHS	Department of Homeland Security
DOJ	U.S. Department of Justice
DOT	U.S. Department of Transportation
e.g.	For Example
EMS	Emergency Medical Services
EMT	Emergency Medical Technician
ERSI	Emergency Responder Safety Institute
ETO	Emergency Transportation Operations
EVSI	Emergency Vehicle Safety Initiative
FGC	Fireground Command
FHWA	Federal Highway Administration
FIRESCOPE	FIre RESources of California Organized for Potential Emergencies
ft	Feet
GM	General Motors
GPS	Global Positioning System

HRHIMP	Hampton Roads Highway Incident Management Plan
HSPD	Homeland Security Presidential Directive
IAFC	International Association of Fire Chiefs
IAFF	International Association of Fire Fighters
IAP	Incident Action Plan
IC	Incident Commander
ICP	Incident Command Post
ICS	Incident Command System
ID	Identification
IFSTA	International Fire Service Training Association
IMS	Incident Management System
IMT	Incident Management Team
ISEA	International Safety Equipment Association
ITS	Intelligent Transportation Systems
JPO	Joint Program Office
LED	Light Emitting Diode
LW	Lane Width
MOU	Memorandum of Understanding
MPH	Miles Per Hour
MUTCD	Manual of Uniform Traffic Control Devices for Streets and Highways
N/A	Not Applicable
NCHRP	National Cooperative Highway Research Program
NFA	National Fire Academy
NFPA	National Fire Protection Association
NFSIMSC	National Fire Service Incident Management System Consortium (now known as the National Incident Management System Consortium; NIMSC)
NHTSA	National Highway Transportation Safety Administration
NIC	NIMS Integration Center
NIJ	National Institute of Justice
NIMS	National Incident Management System

NIMSC	National Incident Management System Consortium (formerly known as the National Fire Service Incident Management System Consortium; NFSIMSC)
NIOSH	National Institute for Occupational Safety and Health
NIST	National Institute of Standards and Technology
NTIMC	National Traffic Incident Management Coalition
NRF	National Response Framework
NUG	National Unified Goal
NVFC	National Volunteer Fire Council
OSU	Oklahoma State University
PIO	Public Information Officer
POV	Privately Owned Vehicle
PPE	Personal Protective Equipment
PSA	Public Service Announcement
PSAP	Public Safety Answering Point
PSE	Planning for Special Events
PSL	Posted Speed Limit
SAE	Society of Automotive Engineers
SCBA	Self-Contained Breathing Apparatus
SOG	Standard Operating Guideline
SOP	Standard Operating Procedure
SUV	Sport Utility Vehicle
TIM	Traffic Incident Management
TIMA	Traffic Incident Management Area
TIMS	Traffic Incident Management Systems
TL	Taper Length
TMS	Traffic Management System
TSP	Telematics System Providers
TTC	Temporary Traffic Control
UC	Unified Command
UMTRI	University of Michigan Transportation Research Institute
USFA	U.S. Fire Administration

UV	Ultraviolet
VDOT	Virginia Department of Transportation
VII	Vehicle Infrastructure Integration
VSP	Virginia State Police

Appendix B Resource Websites and Information Sources

The following websites and information sources contained useful information on traffic incident management (TIM) and roadway incident management safety at the time this report was written. Website addresses do change on occasion and some websites are discontinued, so each of these site's availability cannot be ensured in the future.

Ambulance Visibilty
This website provides information on international practices for increasing the visibility of emergency medical services (EMS) vehicles. http://ambulancevisibility.com

"Alive on Arrival"
2010 U.S. Fire Administration (USFA) publication featuring safe tips on emergency-vehicle response. www.usfa.fema.gov/downloads/pdf/publications/fa_255f.pdf

Battenburg Markings on Emergency Vehicles
Information on Battenburg markings for emergency vehicles. http://en.wikipedia.org/wiki/Battenburg_markings

Drive to Survive Website
This website has safety information on emergency-vehicle safety. www.drivetosurvive.org

"Effects of Warning Lamp Color and Intensity on Driver Vision"
A 2008 USFA/Society of Automotive Engineers (SAE) report on this topic. www.sae.org/standardsdev/tsb/cooperative/warninglamp0810.pdf

"Effects of Warning Lamps on Pedestrian Visibility and Driver Behavior"
A 2008 USFA/SAE report on this topic. www.sae.org/standardsdev/tsb/cooperative/nblighting.pdf

Emergency Responder Safety Institute (ERSI)
Their main website is at: www.respondersafety.com

Their "Highway Incident Safety for First Responders" PowerPoint® training program may be downloaded at: www.lionvillefire.org/hwy_safety

Federal Highway Administration (FHWA; U.S. Department of Transportation (DOT))
Numerous resources are provided by the FHWA at the following websites:

Traffic Incident Management Handbook.
http://ops.fhwa.dot.gov/eto_tim_pse/publications/timhandbook/tim_handbook.pdf

"Best Practices in Traffic Incident Management."
http://ops.fhwa.dot.gov/publications/fhwahop10050/index.htm

Manual on Uniform Traffic Control Devices for Streets and Highways.
http://mutcd.fhwa.dot.gov

"Simplified Guide to the Incident Management System for Transportation Officials."
http://ops.fhwa.dot.gov/publications/ics_guide/ics_guide.pdf

Firefighter Close Calls
This website contains news and other information related to all aspects of firefighter safety. www.firefighterclosecalls.com

"Hampton Roads Highway Incident Management (HIM) Regional Concept for Transportation Operations (RCTO)"
This 2008 document may be downloaded at: www.hrtpo.org/Documents/Reports/2008/RCTOExecSummFinal%20Copy.pdf

I-95 Corridor Coalition
Upload the I-95 Corridor Coalition's "Coordinated Incident Management Toolkit for Quick Clearance" at: www.i95coalition.net/i95/Portals/0/Public_Files/uploaded/Incident-toolkit/toolkit_document_dvd.pdf

International Association of Chiefs of Police
The Arizona Blue Ribbon report on police vehicle safety. www.theiacp.org/Portals/0/ppts/AZ_DPS/AZ_DPS_files/frame.htm

International Association of Fire Chiefs (IAFC) *Guide to Model Procedures for Emergency Vehicle Safety*
This guide can be downloaded for free from the following website: www.iafc.org/vehiclesafety

International Association of Fire Fighters (IAFF) Response and Roadway Safety Program
This program can be downloaded for free from the following website: www.iaff.org/hs/evsp/home.html

International Association of Fire Fighters (IAFF) Best Practices for Emergency Vehicle and Roadway Operations Safety in the Emergency Services
This program can be downloaded for free from the following website: www.iaff.org/hs/EVSP/guides.html

***Manual on Uniform Traffic Control Devices for Streets and Highways* (MUTCD)**
This document can viewed online or downloaded for free at: http://mutcd.fhwa.dot.gov

Minnesota Traffic Incident Management Recommended Operational Guidelines
Their main website is located at: www.dot.state.mn.us/tmc/documents/Freeway%20Incident%20Management.pdf

National Firefighter Near-Miss Reporting System
This site allows firefighters to report and search reports on near-miss safety incidents: www.firefighternearmiss.com

National Fire Protection Association (NFPA)
Their various standards that apply to vehicle and roadway safety can be previewed for free at: www.nfpa.org

National Highway Traffic Safety Administration (NHTSA)
Their main website is at: www.nhtsa.dot.gov

National Incident Management System Consortium (NIMSC)
Their main website is at: www.ims-consortium.org

Information on their publications titled *Incident Command System Model Procedures Guide for Incident Involving Structural Fire Fighting, High Rise, Multi-Casualty, Highway and Managing Large-Scale Incidents using NIMS-ICS and IMS Model Procedures Guide for Highway Incidents* can found at www.ifsta.org or by calling (800) 654-4055.

National Institute for Occupational Safety and Health (NIOSH)
The website for their Fire Fighter Fatality Investigation and Prevention Program is at: www.cdc.gov/niosh/fire

The National Law Enforcement Officers Memorial Fund
Their mission is to generate increased public support for the law enforcement profession by permanently recording and appropriately commemorating the service and sacrifice of law enforcement officers; and to provide information that will help promote law enforcement safety. www.nleomf.com

National Safety Council (NSC)
Online defensive driving courses and information available from the National Safety Council. www.nsc.org/ddc/training/ddconline_train_courses.aspx

National Traffic Incident Management Coalition (NTIMC)
Their main website is at: http://timcoalition.org/?siteid=41

North Florida Transportation Planning Organization (TPO) TIMe4Safety Program
This program includes a handbook and video presentations. www.northfloridatpo.com/index.php?id=25

Nova Scotia Traffic Management Guidelines for Emergency Scenes
Their main website is at: www.gov.ns.ca/lwd/firesafety/docs/EmergencyRespondersTrafficManagement Guidelines-EmergencyScenes.pdf

National Volunteer Fire Council (NVFC) Emergency Vehicle Safe Operations for Volunteer and Small Combination Emergency Service Organizations
This program can be downloaded for free from the following website: www.nvfc.org/evsp/index.html

State of New Hampshire Memorandum of Understanding (MOU) for Statewide Traffic Incident Management
This example of a Statewide MOU can be downloaded at: www.i95coalition.org/i95/Portals/0/Public_Files/uploaded/Incident-toolkit/documents/MOU/ MOU_QC_NH.pdf

The Officer Down Memorial Page
This page provide statistics and case study information on police officer fatalities. www.odmp.org

Police Driving.com
This site is dedicated solely to improving the safety of driving police vehicles. www.policedriving.com

State of Tennessee "Strategic Plan for Highway Incident Management in Tennessee"
This document outlines a Statewide plan for highway incident management. www.tdot.state.tn.us/incident/CompleteIMPlan.pdf

U.S. Fire Administration (USFA)
The USFA website is at: www.usfa.fema.gov

The USFA Roadway Operations Safety website is at: www.usfa.fema.gov/fireservice/research/safety/roadway.shtm

The USFA Emergency Vehicle Safety website is at: www.usfa.fema.gov/fireservice/research/safety/vehicle.shtm

U.S. Department of Justice (DOJ)
Download the report titled "Evaluation of Chemical and Electric Flares" at: www.ncjrs.gov/pdffiles1/nij/grants/224277.pdf

U.S. DOT Emergency Transportation Operations
The U.S. DOT FHWA website on handling roadway emergencies. http://ops.fhwa.dot.gov/eto_tim_pse/index.htm

U.S. DOT Intelligent Transportation Systems Project
Their main website is at: www.its.dot.gov/index.htm

VFIS
VFIS has emergency vehicle driver and instructor materials available. www.VFIS.com

www.ingramcontent.com/pod-product-compliance
Lightning Source LLC
Chambersburg PA
CBHW081136170526
45165CB00008B/2698